Secrets to Great Soil

A Grower's Guide to Composting, Mulching, and Creating Healthy, Fertile Soil for Your Garden and Lawn

ELIZABETH STELL

The mission of Storey Communications is to serve our customers by publishing practical information that encourages personal independence in harmony with the environment.

Edited by Carol M. Healy and Gwen W. Steege
Cover and text design by Mark Tomasi
Cover photograph by Dwight R. Kuhn Photography
Production assistance by Susan Bernier and Eileen Clawson
Illustrated by Alison Kolesar and Susan Berry Langsten, except page 17 by Cathy Baker; pages 36 (top), 38 (bottom), 39 (middle right), 40 (top), 42 (top), 46, and 48 (left) by Brigita Fuhrmann; and pages viii, 36 (lower left), and 81 (middle) by Elayne Sears
Indexed by Northwind Editorial Services

Printed in Canada by Transcontinental Printing
10 9 8 7 6 5 4 3 2

Library of Congress Cataloging-in-Publication Data

Stell, Elizabeth.
Secrets to great soil / Elizabeth Stell.
p. cm. — (Storey's gardening skills illustrated)

Includes bibliographical references and index.
ISBN 1-58017-009-9 (hardcover : alk. paper). — ISBN 1-58017-008-0 (pbk. : alk. paper)
1. Garden soils. 2. Soil management. 3. Gardening. I. Title. II. Series.
S596.75.S74 1998
635'.0489—dc20
97-40932
CIP

Acknowledgments

I'd like to thank Dr. Rich Koenig of the Utah Cooperative Extension Service, Garn A. Wallace of California's Wallace Laboratories, Steve Bodine of the University of Massachusetts Soil and Plant Tissue Testing Laboratory, and Jonathan Collinson of Woods End Research Laboratory in Maine for assistance with technical questions; also Naka Ishii of the University of Massachusetts Morrill Biological Sciences Library for research assistance.

Dedication

to Bill
who makes everything possible
(and more fun)

SECRETS TO GREAT SOIL

Contents

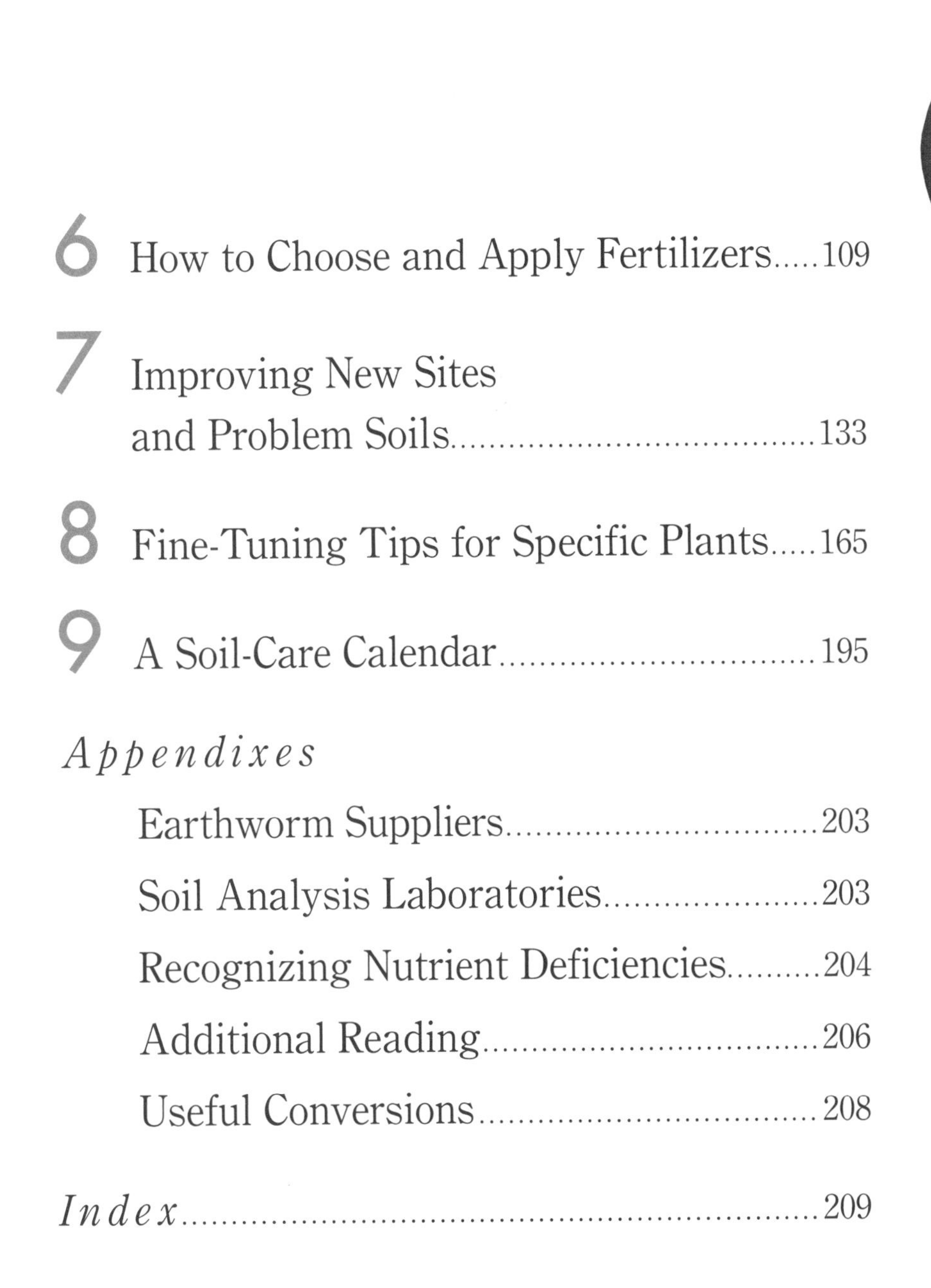

Introduction

I've never outgrown a child's simple love of dirt. It has continued to fascinate me since those early days of mud pies and simple earthworks. After studying it at the university level I learned to respect its complexity and call it soil. Years of gardening have taught me how resilient it is. The more I learn about soil, the more marvelous and magical it seems.

My first attempts at soil improvement were accidental but astonishingly successful. In a fit of environmental enthusiasm in my early teens, I decided our tidy suburban house needed a compost bin. My parents weren't exactly receptive. They vetoed a bin but were persuaded to let me try trench composting. I wasn't even thinking of the soil, just of recycling kitchen scraps.

With great enthusiasm but weak arms and a dull shovel, I started my project. No luck. Our Maryland clay soil was baked so hard by summer sun that the shovel simply bounced off! I had no idea soil could be so hard. (I had yet to see the adobe-like soils of the Southwest!) Later, after softening rains, I tried again with better results. I dutifully emptied kitchen scraps into the trench, covering each addition with a bit of soil.

I filled the trench before winter and forgot about it until spring. When I dug around to see what had happened, I couldn't believe my eyes. Loose, crumbly, easy-to-dig soil with lots of earthworms I hadn't seen before! The dark soil between my fingers seemed completely unrelated to the light-colored stuff that had resisted my shovel only a few months before. My parents were impressed, too. The conclusion was simple: Soil is magical stuff, yet the alchemy for transforming stubborn clay into lovely loam is surprisingly straightforward.

The Fertile Place

"We all live lives full of compromise: some work day jobs sitting behind desks, secretly waiting to come home to renew ourselves in the garden, others have made gardening our livelihood but find the demands of a business have distanced us from hoe and spade. It is a balancing act, we tell ourselves, but how do we achieve and maintain that balance? The earth is a grounding force in our technological lives, our gardens havens from the ever-faster pace. That which we create around us is a mirror of who we are inside, and the fertile place for our minds to grow."

—Michael Ableman,
National Gardening, 1997

Rescuing Soil on a Larger Scale

About the same time, I discovered a copy of *Malabar Farm* on my grandparents' bookshelf. Louis Bromfield reached a conclusion similar to mine from efforts on a much larger scale in the 1940s. He bought an Ohio farm whose soil was so badly eroded that most topsoil was completely gone or marred by large gullies. Though once productive, the farm had been abandoned. After years of failure to replenish organic matter and mismanaged fertility it could no longer produce decent crops.

Bromfield transformed his fields by incorporating as much organic matter as possible as rapidly as possible. He built up the soil with animal manures and green manures, crop rotations that included pasture, and judicious use of lime and synthetic fertilizers. He controlled erosion on sloping fields by growing cover crops rather than leaving the soil bare over the winter and by plowing along the contour rather than straight up and down. He grew strips of sod between strips of easily eroded crops such as corn.

Improvements were visible after only a year or two. Yields increased greatly: Corn yields doubled or tripled in four years, and on some fields wheat yields increased almost tenfold. Every year, fewer pests and diseases bothered the field and garden crops. By the fifth year, insecticides were no longer needed even though an occasional pest was still seen. The animals, fed directly from the farm on pasture, silage, or field crops, became noticeably healthier. (A laboratory analysis of the farm's alfalfa showed it was especially high in nutritious minerals.) A nearby stream muddied with soil washed from the fields became clear again.

During the drought of 1944, when farmers all over Ohio were hauling water, the springs on Malabar Farm were still flowing, because all of the erosion controls had allowed rain to seep into the soil and recharge long-term water reserves.

THE POWER OF TRANSFORMATION

"There is no satisfaction like watching the earth grow richer because of what you do with it. . . . The change in the very landscape from one of abandoned fields, of gullied desolation of hills brown and red with sorrel and broom sedge to green has been as remarkable as the steady darkening color of the soil as the fertility rose with gains in production ranging from 50 to 1500 percent per acre."

— Louis Bromfield, *Malabar Farm,* 1947

Why Should You Become a Soil Steward?

The easiest, most dramatic way to improve any garden is to improve its soil. Efforts to enhance your soil will give you quick results in your gardens, as well as long-lasting benefits. As you build up your soil, it will become crumbly and easy to dig. You'll be rewarded with healthy plants that look better and produce better, even when subjected to weather quirks such as droughts or cold spells.

You can also significantly reduce your pest problems — and your use of pesticides — by building up your soil. Healthy soil just does a better job of producing healthy plants. Vigorous, healthy plants have a greater ability to fight off pests and infections, just as you have an easier time fighting off a cold or flu when you're in excellent health. Fewer insects and diseases in the garden means you handle fewer pesticides and end up with fewer pesticide residues on your garden vegetables and fruits.

Good Nutrition

Feeding your plants well will improve your own diet. Conversely, growing fruits and vegetables on depleted soils reduces their nutrient content.

- Soils low in the minerals required by humans produce fruits and vegetables with low mineral content.
- Soils with too little phosphorus, potassium, or magnesium reduce vitamin C content.
- Soils with too little nitrogen, iron, copper, or molybdenum reduce vitamin A (beta-carotene) content.

Soil improvement benefits the environment in other ways, too. Recycle garden and kitchen wastes into compost; you'll reduce your trash at the same time you reduce your need to buy fertilizer. Healthy soil needs less irrigation water to nurture plants, so you can cut back on water use.

Tests have shown that organically grown produce contains more nutritious minerals (and less sodium) than produce grown by "traditional" methods. Traditional agriculture can get high yields by using lots of fast-acting N-P-K fertilizer even where the soil isn't so great. That's because it resorts to a wide array of pesticides to control any resulting pest problems. The quality of the produce suffers in invisible ways, though. Even without considering pesticide residues, the high yields on mediocre soil are achieved at the expense of mineral content and sometimes even protein quality. For instance, experiments have shown that the insecticide parathion greatly reduces the vitamin C (ascorbic acid) content of spinach. And according to U.S. Department of Agriculture soil scientist Sharon B. Hornick, grains such as wheat and barley grown with high levels of synthetic fertilizers may contain plenty of protein, but it's not good quality protein. In fact, the amino acids are out of balance, so humans can't make use of all of it. High levels of nitrogen fertilizers have also been shown to reduce levels of vitamin C in vegetables such as kale.

Secrets to Great Soil

Soil is surprisingly complex, but improving it is surprisingly simple. As I learned early on, dense clay can be transformed into crumbly loam simply by burying kitchen scraps and waiting. I've even converted a gravel drive into a productive vegetable bed. A few years ago, I realized that a former driveway had the best sun exposure in our tree-lined yard. A dairy farmer friend supplied a truckload of ancient manure and my husband mixed this into the packed gravel with our ancient rototiller. I sifted out the gravel from a few cartloads but soon lost patience to sift any more. Our new vegetable garden had more sand and gravel than my other beds — but then I'm used to stone after gardening on a New England hillside. While I grow my carrots elsewhere, I grow great potatoes, onions, and other vegetables on this former driveway.

Your soil probably isn't as extreme as baked clay or a gravel driveway. Using this book, you can improve any soil from coastal Californian sand to Southeastern clay. *Secrets to Great Soil* shows you how to play detective to uncover what you've got and why your soil is the way it is. Learn your NPKs, and how to choose the best fertilizer to supply these and other nutrients. Discover tools to make your work easier. Find out how many different ways there are to make compost, plus other easy options for supplying organic matter — the cure-all for soils. Learn strategies for problem sites and tips for tailoring any soil to suit your needs, from lawns to trees to container gardens. Seasonal reminders are included in a handy soil-care calendar. Whether you're just starting out or a seasoned gardener, you'll find what you need here to nurture your patch of earth.

CHAPTER·1

Get to Know Your Soil

Gardeners are an idealistic bunch, so it's only natural that they try for ideal soil. The soil experienced gardeners crave is deep and easy to work. It allows plants to push through easily and develop many long roots. A handful of ideal soil has a nice, crumbly feel and a chocolate color. It smells earthy and rich, even sweet. Such soil is called loam.

Loam, Sweet Loam

Technically *loam* refers to a specific, well-balanced soil texture. But it's commonly used more loosely to mean ideal soil. That's not surprising, as both meanings of the word are similar in terms of physical characteristics. These physical qualities work together to promote healthy plant growth. When one is out of line, other physical characteristics may suffer. Plant growth suffers, too. Of course, ideal soil is also fertile soil; when people use loam in its looser sense, they imply good fertility as well.

The balanced texture of loam means it retains a good amount of water and nutrients. It doesn't dry out too fast after it rains, yet it doesn't stay soggy. Loam's well-developed structure resists compaction and reduces the risk of erosion. Such soil keeps its small crumbs when you dig it and when thunderstorms pound it. Good structure in turn ensures good drainage, so long-lasting puddles never form. Loam contains lots of organic matter, which in turn enhances all of the other physical characteristics. Organic matter also ensures a healthy soil community with large populations of earthworms and other beneficial organisms.

Soil building consists of making your soil as much like loam as possible or, if you're blessed with good loam, maintaining its productive qualities. That means fixing those characteristics that are less than ideal but easy to change, such as drainage. Soil building also means learning how to manage (or compensate for) the characteristics you can't change, such as texture.

In This Chapter

- What Is Soil?
- How Your Soil Was Formed
- Reading the Layers
- Sand, Silt, or Clay?
- Testing Soil Texture
- The Structure of Soil
- Examining Soil Structure
- Promoting Drainage
- Encouraging Soil Life
- The Industrious Earthworm

DID YOU KNOW?

Tilth

In addition to loam, there's another word that describes the structure of soil: *tilth.* It comes from an ancient word for cultivation, as in "to till the soil." Soil with good tilth has all the wonderful characteristics of loam. "Good tilth" means good soil, so it's something to strive for.

What Is Soil?

Though the earth beneath our feet seems solid, only half the soil is hard particles. The other half is a combination of space (air) and water. In good soils, about half of this "empty" half is filled with water. Dissolved in this water are lots of chemicals, including the nutrients essential for plant growth. The remaining space or air is also important. It supplies oxygen to plant roots and soil organisms.

Though you may not realize it, good soils are teeming with a variety of life. Most soil creatures are microscopic, though some, such as worms and ants, are easy to see. Plants (and gardeners) depend on soil organisms to convert nutrients into the forms they can use, and to recycle nutrients from organic matter. The different forms of life present in soil interact to create a complex ecosystem. One of the primary ways to build super soil is to nurture this ecosystem and keep it from getting out of balance.

DID YOU KNOW?

What's in Your Soil?

- Mineral particles of varied sizes
- Pebbles and rocks
- Air
- Water
- Nutrients dissolved in soil water
- Bits of undecomposed leaves and other organic matter
- Well-decomposed, unrecognizable organic matter (humus)
- Many organisms of varied sizes

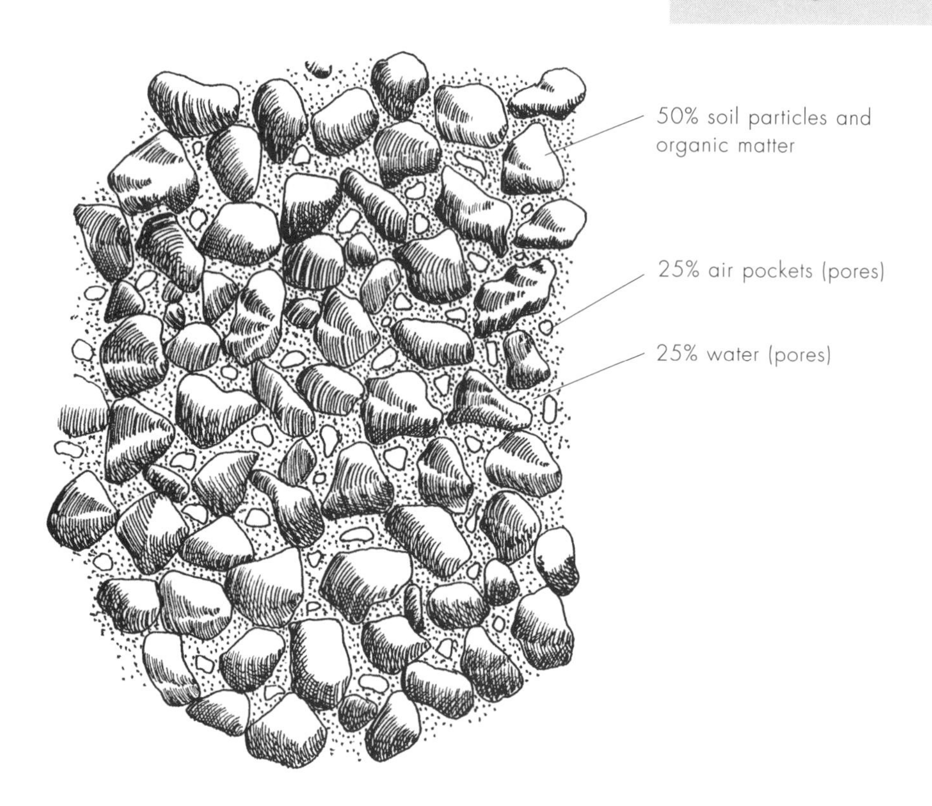

The minute pockets and channels between the solids in soil are called pores. Ideal, loose soil has lots of these and is therefore described as porous. Air can easily enter to reach plant roots, and excess water easily drains away.

The solid portion is mostly tiny bits of pulverized rock. The size of these particles determines the texture — and therefore much of the physical character — of soil. The solid portion also contains bits of dead material in varying stages of decomposition, known as organic matter, which eventually decays into a fine material known as humus.

How Your Soil Was Formed

The nature of the soil in a particular spot is determined by interactions among several factors. Parent material, climate, topography, living creatures, and time influence each other to create soil, as illustrated below.

• **Parent material,** or rocks from which a soil formed, is largely responsible for texture, and sometimes for alkalinity. Sandstones weather into coarse, sandy soils. Soft shales evolve into heavy, clay soils. Granite bedrock usually results in more acidic, sandy loam. Soils derived from limestone are usually fine textured, fertile, and neutral to alkaline.

• **Climate** determines the nature and degree of weathering. Lots of rain or snow greatly speeds up soil formation, but it can also wash away nutrients. Soils in dry climates are often very fertile when first irrigated. But where evaporation rates are high, salts in the soil and irrigation water can concentrate near the surface and reach levels toxic to plants. (Such soils are either saline or sodic; see pages 154–157.) Temperature also affects weathering. Alternate freezing and thawing speeds up the crumbling of rocks. In warm climates, biological activity takes on a much more important role. Warm temperatures also speed up the chemical breakdown of iron from parent rocks, forming the red and yellow soils common in the southern United States and in tropical climates.

• **Topography,** the height and shape of the land surface, influences the depth as well as the texture of soil. Soils are usually deeper on the flat tops and at the bottoms of hills. They start out shallower and need more protection from erosion on slopes and hillsides. In wet climates, low areas can become waterlogged, turning into peat bogs. In dry climates, low spots are apt to become pockets of salty soil.

• **Animals and plants** are the source of organic matter and many nutrients, released either by decay after they die or in animal wastes such as manure. Direct effects of living creatures range from the burrowing of earthworms or moles to the large-scale earthmoving done by humans.

• **Time** allows all the preceding factors to work together. Over eons, rock and sand break down into the smallest clay particles; younger soils often have little clay.

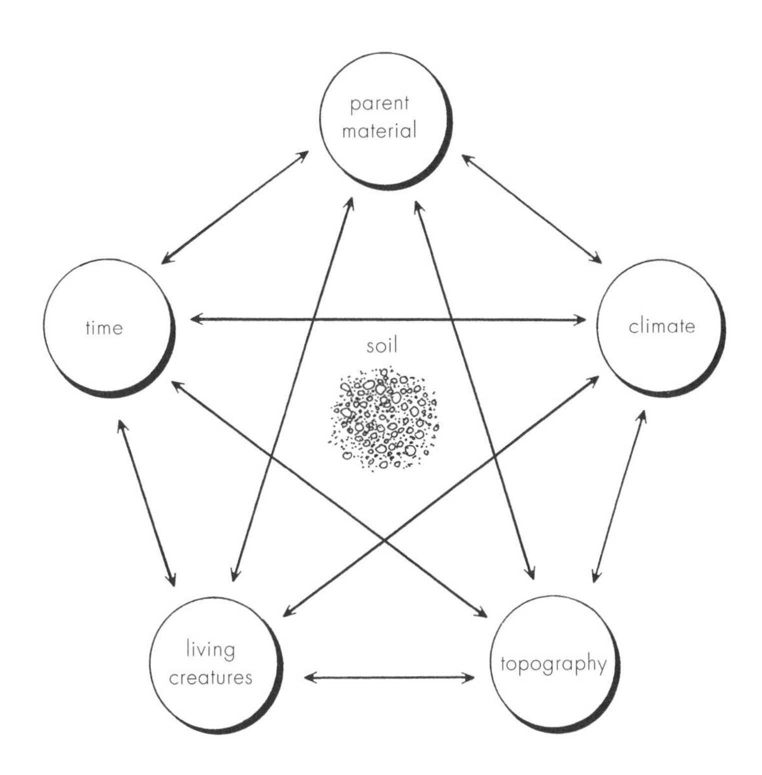

Reading the Layers

As the different soil-forming factors interact over thousands of years, soil matures and forms distinct horizontal layers. The thickness and nature of the layers will vary in different soils. In certain soils, some layers are hard to see or aren't even there. You have to dig 2 to 3 feet deep to see the layer-cake effect.

If you run into bedrock when you dig, your soil is very shallow. It's hard to change the overall depth of soil, at least within a single lifetime. Within garden areas or in raised beds, you'll need to bring in good-quality topsoil. Or you can make lots of compost to add to your existing soil to thicken it.

If you were to dig a hole deep enough, you might encounter groundwater — the water table — before you reached underlying bedrock. If the water table is near the surface for much of the year, it can stop root growth as effectively as a layer of solid rock.

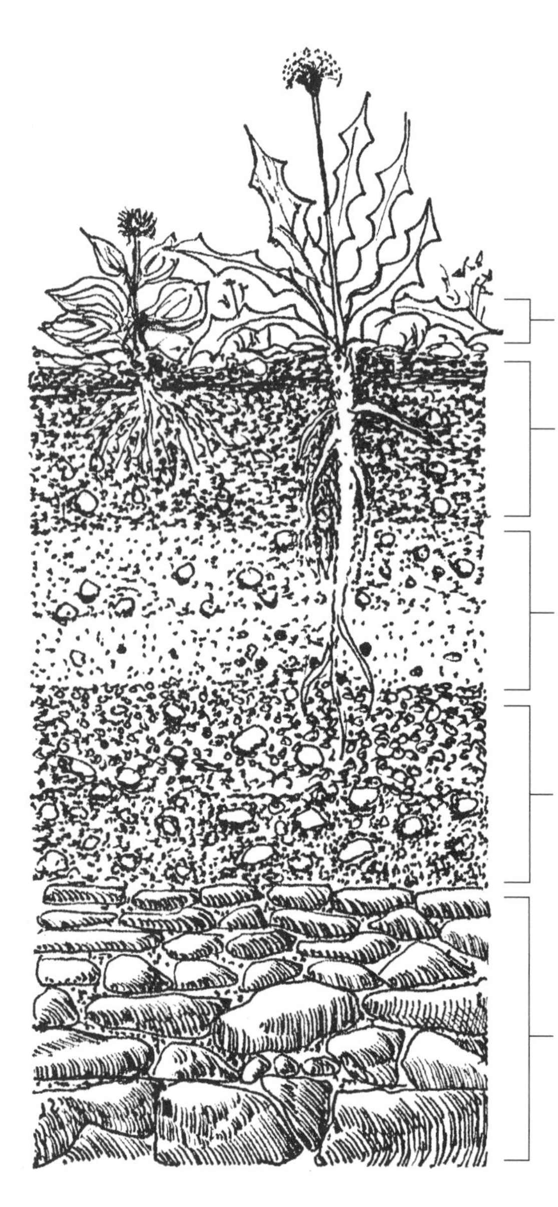

Organic layer. A thin layer of plant material in various stages of decomposition. Absent if the soil has been bulldozed or cultivated.

Topsoil. The top layer of soil, darker and more crumbly than deeper layers, where most nutrients, roots, and soil organisms exist. The deeper this layer, the better.

Subsoil. Usually lighter than topsoil because it contains less humus. Much of the water plants need is stored here, plus some nutrients. Often contains two or more distinct layers. Nutrients, minute specks of humus, and tiny clay particles may be washed into a deeper, darker subsoil layer where they're harder for roots to reach.

Parent material. Rubble that hasn't yet weathered enough to look like soil.

Bedrock. The underlying layer of solid rock, usually too far down to find by digging.

What Your Subsoil Says

Though digging a deep hole takes much effort in most soils, examining subsoil can tell you a lot. Different colors show how well the subsoil drains, which in turn affects the drainage of the overlying topsoil.

What You See	What It Tells You
Red or yellow subsoil	Indicates lots of iron oxides from weathering of parent materials; usually indicates good drainage; often indicates acidic soil; common in warm climates.
Blue or blue-gray subsoil	Indicates lack of oxygen and therefore poor drainage; common in thick layers of clay.
White to ash gray subsoil	Indicates that nutrients and humus have been leached away; usually sits above a darker layer where the leached nutrients and humus have deposited; often indicates acidic and/or sandy soil; common under pines and similar trees.
Even, medium brown subsoil	Adequate drainage.
Pale subsoil with little difference from topsoil	Very young or poorly developed soil; original topsoil may have eroded or been removed (as by bulldozing during house construction).
Dark brown subsoil	Indicates abundant (usually decomposed) organic matter; usually occurs only with peat or muck soils, or where former wetlands have been drained.
Patches or streaks of different colors	Indicates pockets of poor drainage or different soils (see specific colors); plant roots may have trouble moving from one pocket into the next.
Roots all end at same depth	Indicates layer of compacted soil (hardpan) or — in dry climates — a cemented layer (caliche); usually causes poor drainage and hampers plant growth.

Master Gardening Tips

Evaluating Land

Some soils are naturally well suited for gardening; they require the least effort to be tuned into super soil. To see if you have such soil, check the list below. The more your soil differs, the more work it will take to get super soil.

- ***Adequate soil moisture.*** Enough water is available (from rain or irrigation) to support a variety of plants.
- ***Adequate drainage.*** The water table stays below the depth to which most roots grow, and the area isn't flooded more than once every two years.
- ***Adequate soil depth.*** There's enough topsoil over bedrock, hardpan, or gravel to allow for root growth and moisture storage.
- ***Lack of rocks.*** Gravel, stones, or boulders aren't so plentiful that they interfere with tilling or plowing.
- ***Ease of water infiltration.*** The soil should be able to absorb at least .06 inch (.15 cm) of rain or irrigation water per hour in the upper 19.7 inches (50 cm).
- ***Balanced chemistry.*** The soil shouldn't be extremely acidic, alkaline, saline, or sodic.
- ***Even topography.*** Any slopes should be gentle so that the soil doesn't erode easily and so that it's easy to move loaded garden carts or lawn mowers across it.
- ***Moderate temperature.*** The average annual soil temperature should be higher than 32°F (0°C), and the average summer soil temperature should be higher than 46°F (8°C).

Sand, Silt, or Clay?

One of the most important characteristics to know about your soil, texture is simply the relative amounts of different-size mineral particles present. Soils contain a mix of different particle sizes and are described by the particle size that dominates their particular mix. While some people call sandy, silty, or clayey their soil type, they're really describing its texture.

Texture has a major effect on the physical properties of soil. As particle sizes get larger, the spaces between particles get larger, too. This explains why sandy soils drain more quickly than other soils. The smaller the particles, the more surface area they provide for holding water and nutrients. This is why soils containing more clay have a greater ability to hold nutrients and moisture.

As you might guess, there's a trade-off in the advantages offered by different textures. Too much clay results in a soil with plenty of nutrients but problems from poor aeration and slow drainage. Such soils are described as heavy. Too much silt can also cause drainage problems. Too much sand means there's never a drainage problem, but you have to keep adding fertilizer, as most washes away. Such soils are described as light.

The best soil texture isn't made up of all medium-size particles but of a balanced mix of sizes. Loam is the technical word used to describe this ideal texture; such soils are the best ones for gardening and farming. Loams are usually subdivided into sandy loams, clay loams, silt loams, and even silty clay loams.

DID YOU KNOW?

Where the Action Is

The surface of soil particles is where much of the water and nutrient reserves are held. It's also where many important chemical reactions take place. Think of particle surfaces as tiny magnets. More surface area means a greater ability to hold water and nutrients where plants can get them.

Fine clay can provide 10,000 times more surface area than the same weight of medium-size sand grains; 1,000 times more surface area than silt. A pound of pure clay provides up to 90 acres of surface area!

The finest soil particles, those smaller than .0001 inch and too small to see, are the clay particles. Next largest are silt particles, larger than clay but smaller than .002 inch and still too small to see. Pure clay or silt feels mostly smooth between your fingers. Sand particles range in size from .002 to .08 inch. They're usually large enough to see and feel gritty between your fingers. Anything larger than sand is considered gravel.

Common Characteristics of Soils with Different Textures

High Sand Content	High Silt Content	High Clay Content
• Can cultivate when moist or wet • Somewhat dusty when dry • Doesn't form clods • Quick to warm up in spring • Low levels of organic matter • Organic matter breaks down quickly (especially in warm, humid areas) • Easy to correct soil acidity or alkalinity, but correction doesn't last long • Low to moderate risk of water erosion • Feels gritty between fingers	• Harder to cultivate when moist or wet • Often very dusty when dry • Rarely forms clods • Somewhat slow to warm up in spring • Moderate to high levels of organic matter • Organic matter breaks down somewhat quickly • Hard to correct soil acidity or alkalinity • High risk of erosion in windy areas • High risk of erosion by water • Feels silky between fingers (any grittiness is very fine)	• Do not cultivate when wet and sticky (high risk of compaction) • Hard or cementlike when dry • Slow to warm up in spring • High to moderate levels of organic matter • Organic matter breaks down slowly • Hard to correct soil acidity or alkalinity (but once corrected, effects last) • Low risk of erosion in windy areas • Low risk of water erosion if structure is good • High risk of water erosion if structure is poor • Feels smooth between fingers, sticky when wet

Managing Texture

There's a simple fix for soils with less-than-ideal textures: Add organic matter. Lots of organic matter makes soils of any texture behave more like an ideal loam. While adding pure clay to sand will improve its texture, in practice that's hard to do. It's easier to buy coarse sand to add to clay, but it's really hard to add enough to improve drainage. Organic matter is much easier to add — and more effective. Abundant amounts of organic matter increase the ability of sandy soils to stay moist and retain nutrients. Magically, they also improve the aeration and drainage of clay soils. Chapters 4 and 5 explain many ways to increase your soil's organic matter.

Extreme textures may need more than organic matter. Clay soils must be handled carefully and often benefit from double-digging. Building a raised bed may be required where drainage problems are severe. Careful plant choice will help you in very sandy soils — or any other extreme soil. Chapter 7 explains techniques and plants for soils with problem textures.

How Texture Affects Soil Properties

Texture	Aeration/ Porosity	Ease of Water Infiltration	Ability to Hold Nutrients	Water-Holding Capacity	Ease of Working
Loam	medium	medium	medium	medium	medium
Clay	poor	poor	excellent	good	poor
Silt	medium	medium	medium	medium	medium
Sand	excellent	good	very poor	very poor	good

Testing Soil Texture

Classify your soil's texture by measuring the relative amounts of sand, silt, and clay with the following simple test. Since your garden varies slightly from spot to spot, your sample should combine soil from several spots to give you an average. Avoid extreme areas, such as unusually wet or dry spots. (Try testing these areas separately to see if soil texture explains why they're different.)

MATERIALS

- Bucket
- Trowel
- Quart jar with lid
- Water
- Dishwasher detergent
- Clock or watch
- Grease pencil
- Ruler
- Calculator (optional)

1 Dig a 6-inch-deep (15 cm) hole in six different spots throughout your garden. Cut a long slice of soil from the side of each hole. Put all of your slices in the bucket and stir well. You should have at least 2 cups (473 ml) of mixed soil.

2 Place 1 cup (237 ml) of mixed soil into the jar. Remove any large pebbles, sticks, or plant parts. Fill the jar nearly to the top with water. Add about ½ teaspoon (2 ml) of dishwasher detergent, cover, and shake the jar vigorously. If any lumps remain, shake again.

Master Gardening Tip

Classifying Your Results

Check your percentages against those here to figure out your soil's texture. Clay loam is common; if your soil also has a lot of either silt or sand, call it a silty clay loam or a sandy clay loam. Sandy loam is also common; if you have 70–85% sand, call your soil loamy sand.

55% clay = clay soil	85% sand = sandy soil	85% silt = silt soil
20–55% clay = clay loam	50–70% sand = sandy loam	50–80% silt = silt loam

3 Set the jar of muddy water where it won't be jostled for at least 24 hours. After ½ to 1 hour, you should see a distinct layer settled out at the bottom. This is the sand layer. With the grease pencil, draw a line on the jar to mark the top of this layer.

4 Wait another hour. You should now see a second layer; if not, check back in one more hour. This is the silt layer. Mark the top of this layer with the grease pencil.

5 After 24 hours, or when the water is clear, mark the top of the next, clay layer. Don't jostle the jar; clay is very easy to stir up. A quick glance will give you the relative proportions of sand, silt, and clay. To calculate the percentages, measure the height of each layer. Divide the total height into the height of one layer; multiply your result by 100.

Hints for Success

Sample Calculation

Assume you have 2 inches of soil in your jar with a ¼-inch sand layer, a ½-inch silt layer, and a 1¼-inch clay layer. You'll then have

$0.25 \div 2 = .125 \times 100 = 13\%$ sand,

$0.5 \div 2 = .25 \times 100 = 25\%$ silt,

$1.25 \div 2 = .625 \times 100 = 63\%$ clay.

The Structure of Soil

Structure describes how soil hangs together, how much soil particles clump into crumbs or clods. Loose crumbs and clods ensure ample pore space, no matter what your soil texture. As a result, good structure can compensate for less-than-ideal texture. Soil with good structure absorbs more rainfall more quickly, and excess water drains quickly away. Roots and soil organisms push through more easily, and gardeners dig with ease. Good structure makes good gardens.

Organic matter — in partnership with soil organisms — is the main agent behind good structure. It greatly increases the pore space in soil. It not only helps crumbs (aggregates) form, it also increases their stability. Earthworms and microorganisms break down organic matter into gelatinous substances that gently hold soil particles together. Roots of plants and fungi also contribute, in two ways. They push soil particles together as they push through. They also manufacture gummy substances that help hold these particles together.

Managing Structure

As with texture, the easiest way to improve soil structure is by adding organic matter. Once you've achieved good structure, it's impossible to maintain it without maintaining humus levels. Organic matter (plus a few months' time) may be all you need to improve powdery soil or large, hard clods. Compacted structure, soil that's crusted on the surface or forms a dense layer (hardpan) below the surface, is harder on plants and takes more effort to fix. Rainwater and roots can't enter easily, and it can be hard to dig. Adding organic matter isn't enough; you'll also need to try double-digging (see page 141) or loosening with a broadfork (see page 143).

When to Work Your Soil

A simple squeeze tells if your soil is ready for digging or cultivating. Grab a handful of soil and squeeze it. When you open your hand, do you see just loose powder? Your soil is too dry. If you can't wait for rain, run a sprinkler long enough to moisten the top few inches and wait several hours (ideally, overnight) before digging. Does water run out when you squeeze, or do you see a solid, sticky lump in your hand? Your soil's too wet. Wait a day or so and test another handful before digging. Do you see a mostly solid lump that easily breaks apart if you poke it? Dig away! Your soil is slightly moist, just right for cultivating.

HINTS FOR SUCCESS

How to Promote Ideal Soil Structure

- Keep levels of organic matter high: Add compost, manure, or plant residues regularly, or grow green manure crops.
- Encourage abundant, healthy soil organisms (by adding organic matter).
- Cultivate soil only when moist: Avoid walking on, or driving machines over, wet soil.
- Make sure soil contains a good balance of calcium and magnesium (but avoid overliming).
- Cover bare soil: Mulch areas that aren't covered by plant leaves to minimize pounding from heavy rains.
- Keep soil in use as much as possible: Plant a green manure or cover crop if an area will be bare for a couple of months.
- Minimize rototilling and other forms of cultivation, as these break up desirable soil crumbs and channels.

Examining Soil Structure

The best way to learn about the structure of your soil is to look at it. Start by digging a hole and comparing the soil in place on the sides with the loose soil removed from the hole. To test the stability of your soil's structure, see if its crumbs hold up when you add water.

1 Dig a hole about a foot (30 cm) deep. Can you see small (¼- to ½- inch, or .5 to 1 cm) crumbs in the top few inches? Ideally, you'll see some crumbs even 12 inches (30 cm) down. A crust on the surface, or a dense layer (hardpan) partway down, is a sign of poor structure.

2 Does the soil removed from the hole contain lots of ½-inch (1 cm) crumbs? Do larger clods break apart easily? These are signs of good structure.

3 Place a handful of soil crumbs in each of two glasses. Gently pour water into one of the glasses to cover the soil. If your soil has good (stable) structure, crumbs will hold together even when wet, so the wet and dry samples should look similar.

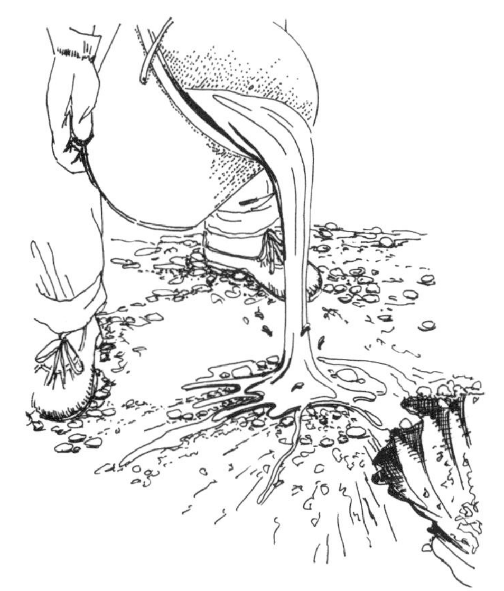

4 Pour water onto bare soil in the garden. Does it flatten all the crumbs into a fine, even layer? This effect, called puddling, is a sign that your structure needs improving.

Promoting Drainage

How quickly water moves through your soil — drainage — affects how plants grow. Very poor drainage suffocates plants. When soil is waterlogged, all the pore spaces that normally hold air are instead filled with water, so roots can't get oxygen. Oxygen-starved soil also suffocates earthworms and other beneficial soil organisms.

At the opposite extreme, soils that drain too quickly can't supply roots with enough water. Such soils also don't hold onto nutrients long enough for plants to use them. Gardeners with excessively drained soil find that they have to irrigate and fertilize constantly.

The Importance of Structure and Texture

Good structure helps ensure good drainage. Channels between crumbs allow water to move through and drain away so that it doesn't collect. The sticky substances (humus or clay) holding crumbs together also help hold water from draining away too quickly. Texture can help or hurt. The sandier your soil's texture, the more quickly it will drain; the more clay in your soil, the slower it will drain. The universal soil improver — organic matter — helps correct any drainage problems by balancing texture and improving structure.

Problems Underground

Conditions below the surface can either help or complicate matters. A heavy clay soil sitting on a thick layer of sand or gravel may never experience drainage problems. An ideal, loamy soil sitting on solid clay or compacted soil (hardpan) may collect large puddles after heavy rains. Even if your soil's texture and structure are great, you can still have waterlogging from a high water table (where groundwater is close to the surface).

MASTER GARDENING TIP

Dealing with Topography

The shape of the land surface, or topography, has a big effect on drainage. Where ground slopes or forms a small hill, drainage of heavy-textured soils will improve. Slightly sandy soils, though, may behave more like very sandy soils. In low spots or where land forms a bowl, water will tend to collect regardless of how sandy the soil is.

You can modify the topography to correct all but the most extreme drainage problems. In low spots, in heavy soils, or where the water table is high, building raised beds will improve drainage. In sandy (light) or gravelly soils and dry climates, creating sunken beds will help gardens capture all available rain or irrigation water. (See page 145 for instructions.) Where topography causes severe problems, get advice from a landscaping professional about regrading your yard.

SOIL DRAINAGE CLASSES

Class	Description
Excessively drained	Sandy or gravelly (or on steeply sloping land); top few inches dry out quickly
Well drained	Loamy; top few inches stay moist after rains
Poorly drained	Soil is wet for several months of the year
Very poorly drained	Soil stays wet almost constantly; water forms long-lasting puddles on surface

Plants That Indicate Poor Drainage

Getting to know wild plants in your yard can tell you a lot about what's going on below the surface. The most reliable indicator plants are those for poorly drained soils. While areas that form puddles may be easy to spot after heavy rains, indicator plants can show you these areas at any time. Don't rely on a single plant, or on unhealthy specimens. You need to see several thriving plants of one type, or at least three different types, to know your drainage is poor.

Wildflowers and Weeds

Bulrushes (*Scirpus* spp.)
Buttercups (*Ranunculus* spp.)
Cardinal flower *(Lobelia cardinalis)*
Cattails (*Typha* spp.)
Coltsfoot *(Tussilago farfara)*
Docks (*Rumex* spp.)
False hellebores (*Veratrum* spp.)
Horsetails (*Equisetum* spp.)
Ironweed *(Vernonia noveboracensis)*
Joe-Pye weeds (*Eupatorium* spp.)
Mosses (many spp.)
Plantains (*Plantago* spp.)
Poison hemlock *(Conium maculatum)*
Ragged-robin *(Lychnis flos-cuculi)*
Rushes (*Juncus* spp.)
Sedges (*Carex* and *Cyperus* spp.)
Skunk cabbages (*Symplocarpus* spp.)
Smartweeds (*Polygonum* spp.)
Water hemlocks (*Cicuta* spp.)

Shrubs and Trees

Alders (*Alnus* spp.)
Buttonbush *(Cephalanthus occidentalis)*
Chokeberries (*Aronia* spp.)
Meadowsweets (*Spiraea* spp.)
Summersweet *(Clethra alnifolia)*
Willows (*Salix* spp.)
Winterberry *(Ilex verticillata* and *I. laevigata)*

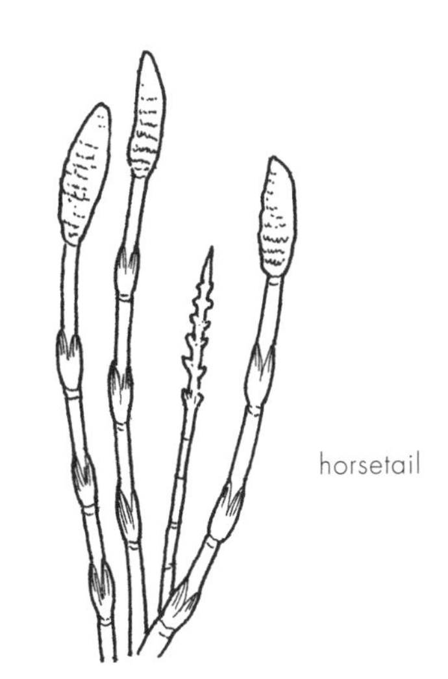
horsetail

poison hemlock

plantain

Testing Soil Drainage

It's easy to test whether your garden has drainage problems even though your texture test shows a good mix of sand, silt, and clay. The following steps are a small-scale version of the "perc" tests used by homeowners in rural areas to see if their yards drain well enough to support a septic system.

For the most accurate results, try this test when the soil isn't extremely wet or dry. Use the squeeze test (see page 10) to make sure your soil doesn't form a sticky, solid lump. Also, try to choose an average spot in your garden. Later, you may wish to test areas where you suspect drainage problems.

1 Dig two holes, each about a foot (30 cm) deep and a foot across, spaced several feet apart.

2 Fill one hole to the top with water. Measure the depth of the water and write it down, noting the time. Repeat with the second hole.

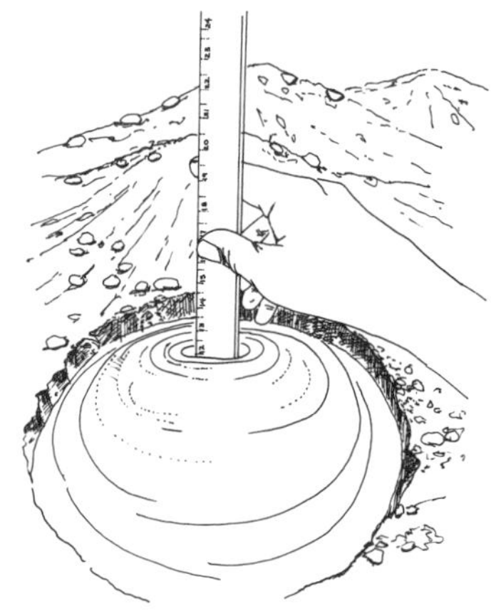

3 After an hour, write down depth and time for both holes. Repeat after two and three hours. Calculate the average inches (cm) lost per hour: Add the six numbers for inches (cm) lost per hour (three from each hole) and divide by six. Water in well-drained soil should drain about an inch (2.5 cm) per hour. If your water disappeared much faster, your soil is excessively drained.

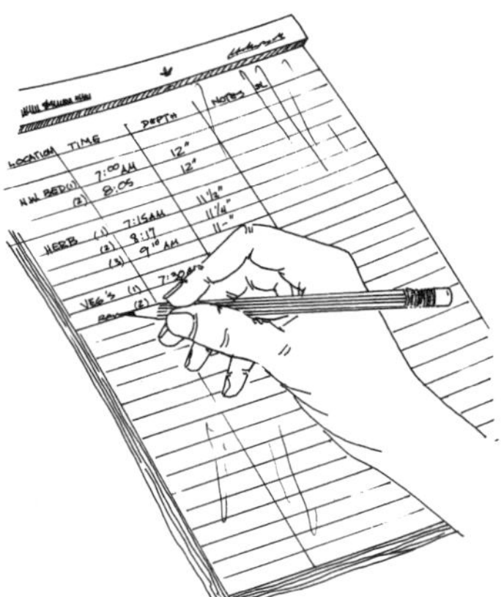

4 Continue to check each hole periodically, and write down the time when all water has drained away. If it takes more than eight hours, your soil is poorly drained.

Life in the Soil

Soil that teems with life is healthy soil. Organisms from earthworms to microscopic bacteria are one of the best signs of good tilth and fertile soil. They need the same soil conditions as plants do to thrive and become numerous. They also help to create the very physical and chemical conditions that constitute good tilth.

Nature's Recyclers

Soil organisms are nature's great recyclers, but making leaves disappear is only part of the picture. As leaves and other plant debris are broken down, all the nutrients bound up in their cells are released into the soil. Roots absorb these nutrients so plants can use them to manufacture new leaves or fruits or seeds. When animals or humans eat these plants, they convert nutrients for their own uses. Eventually, either in manures or by death and decay, the nutrients are returned to the soil.

Added Benefits

Soil organisms offer other benefits. Many stir the soil as they move around, mixing in organic matter as they eat it or carry bits into their nests. By digging aerating channels and excreting gummy substances that stabilize crumbs, they improve structure. They even help balance soil chemistry: Some help neutralize acidity or alkalinity, while others convert unavailable nutrients into the forms used by plants. Still others create the humus that holds nutrients secure against loss by erosion, yet in a form that's easily available to roots.

Fungi are the most versatile of all decomposers. They make major contributions to improving soil structure and fertility. While a few cause plant diseases, others prey on nematodes (tiny soil-dwelling worms) that attack plant roots. Most feed on decomposing material. Fungi range from microscopic, single-celled yeasts to larger molds and the familiar toadstools.

Though they're the smallest decomposers, microscopic bacteria are by far the most numerous, in total weight as well as numbers. They are very simple forms of life, consisting of only a single cell. Since they divide in half to reproduce, their numbers can increase (or decrease) very rapidly. When fresh organic matter is added to the soil, their populations explode, dropping back to lower numbers once their food source shrinks.

DID YOU KNOW?

What Soil Organisms Do for You

- Promote plant growth
- Improve and stabilize soil structure
- Help maintain correct balance of nutrients
- Convert plant nutrients into readily available forms
- Form mutually beneficial relationships with plant roots, increasing roots' uptake of nutrients
- Help neutralize acidic or alkaline soil
- Convert organic matter into humus
- Release the nitrogen and sulfur bound up in organic matter
- Some fix nitrogen in the air into a form plants can use
- Mix soil, and mix organic matter into soil
- Reduce thatch in lawns
- Produce vitamins and growth hormones that benefit plants
- Break down some toxic compounds in soils

Encouraging Soil Life

To enlist the help of soil organisms in creating and maintaining good tilth, create favorable conditions. As these are the same conditions that promote optimum plant growth, you'll be doing your garden a double favor.

Improve Your Soil

The most beneficial types of soil life prefer evenly moist soil that's well aerated and therefore well drained, so start by improving soil texture and structure and ensuring good drainage. In cultivated soils, food is the main limiting factor. That means you need to keep soil abundantly supplied with organic matter to feed soil life. Correct extreme acidity or alkalinity and keep your soil within the pH range of 6 to 8. Also, make sure your soil is well supplied with calcium, as this nutrient is needed by most soil organisms as well as plants.

Use Pesticides and Fertilizers Wisely

Minimize the use of pesticides. Many of these are toxic to soil life as well as to whatever insects you're trying to control. Avoid overfertilizing, because an excess of some nutrients will upset the soil's chemical balance, making other nutrients completely unavailable. Test your soil periodically to see which nutrients are actually needed, and whether any have built up to excess.

Try to minimize use of concentrated, fast-acting types of fertilizers, especially synthetic sources of nitrogen such as ammonia. Direct contact with ammonia harms earthworms, and very concentrated fertilizers can harm many organisms. Less concentrated fertilizers (those with lower N-P-K numbers) and slow-release formulas are safer for beneficial soil organisms. Most organic fertilizers are less concentrated.

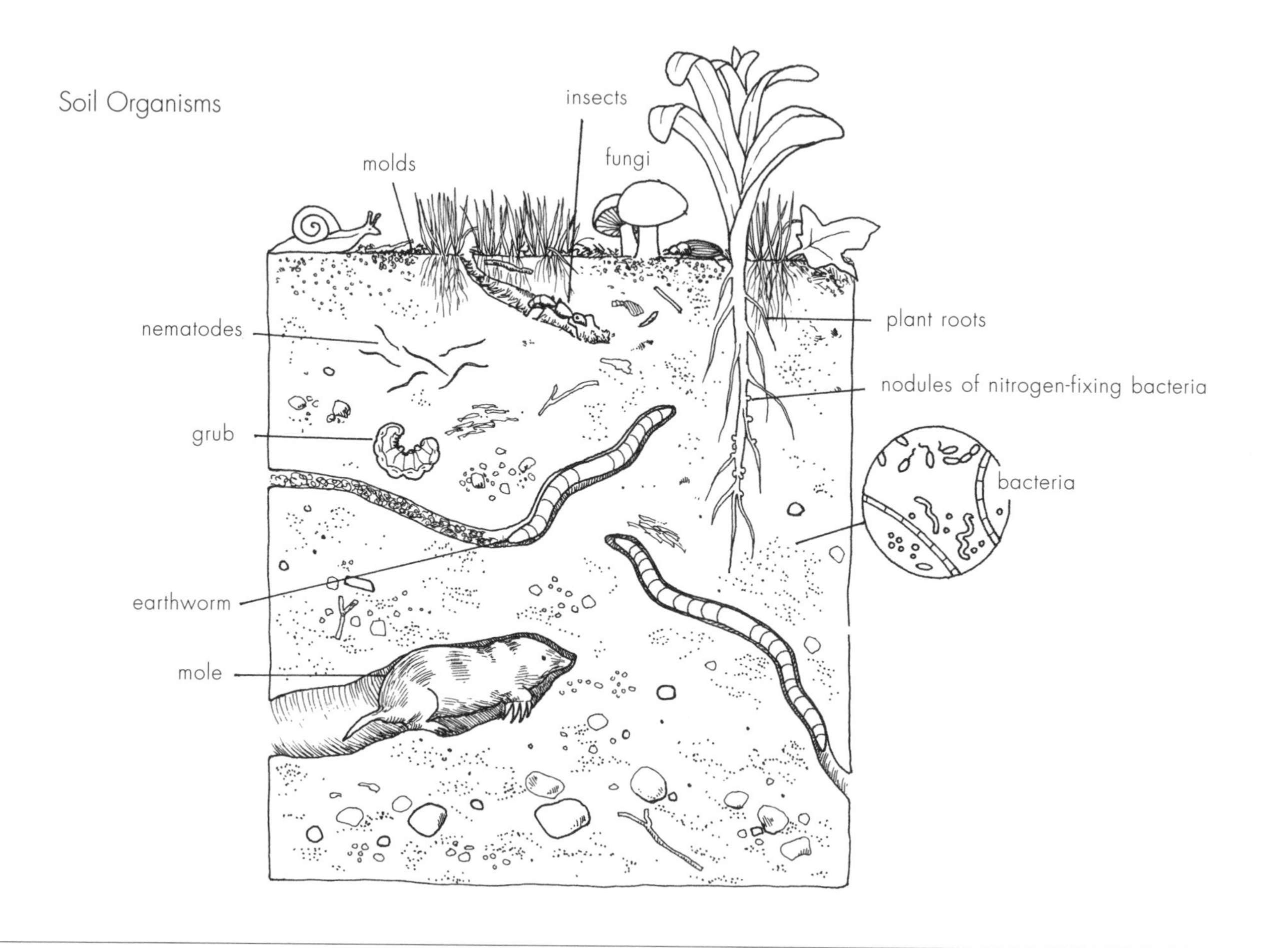

The Industrious Earthworm

In lawns, have you ever noticed small patches of greener grass near small piles of earthworm casts (their wastes or "manure")? Earthworms are the best soil improvers. Studies have shown that an increasing number of earthworms in soil is directly related to increased productivity of plants grown in that soil.

DID YOU KNOW?

Earthworms can swallow and process from 20 to over 200 tons of soil per acre! Where abundant, the soil-enriching casts that earthworms leave on a single cultivated acre can easily amount to 8 tons or more at one time.

Composting Machines

Earthworms provide many of the benefits of other soil organisms. By digging through the earth — as deep as 3 to 6 feet in good soils — they open up channels to increase soil aeration and drainage. They are powerful composting machines, eating large amounts of organic matter and breaking it down into pure humus. Their digestive system provides both a grinding action and enzymes that break down organic matter, mixing it with soil particles (which they can't help swallowing along with bits of debris as they move through soil).

During digestion, earthworms produce gummy substances (including humus) that help hold soil particles together. Earthworm activity increases both the size and the stability of soil crumbs. As a result, these small creatures do more to improve soil structure than any other burrowing organisms.

Casts Are Soil Conditioners

Earthworm casts are an excellent source of nutrients as well as a superb soil conditioner. In addition to being rich in humus, casts are rich in the forms of nutrients most easily used by plants (especially nitrogen, phosphorus, potassium, calcium, and magnesium). The casts usually have a higher pH than the surrounding soil, so they help neutralize acidity. They also increase soil's ability to maintain large, usable stores of nutrients.

DID YOU KNOW?

The common, large red worms known as night crawlers *(Lumbricus terrestris)* aren't native to the United States. They were brought here by early colonists, probably in soil around roots of fruit trees. They're better adapted to surviving in cultivated soil than are many native species of earthworms. As a result, they soon outnumbered native earthworms in cultivated fields and spread quickly across the country wherever soils were cultivated. Native earthworms are still around; they're more numerous in uncultivated soils, such as those under old-growth forests.

An Earthworm Census

Throughout this chapter, the physical benefits of organic matter have been praised. The universal soil improver, it counteracts a difficult texture and helps ensure good structure. It also provides chemical benefits.

Examining the biological activity in your soil is one way to determine if you've got enough organic matter. Earthworms are the easiest members of the soil's biological community to observe. Counting the number of earthworms present in a sample of your soil gives a pretty good idea of its organic matter content and overall health.

1 Choose a 1-foot-square (30 cm^2) site that is a good average of your garden. Dig out the top 6 inches (15 cm) of soil, and place it in a shallow pan.

2 Count the number of earthworms in the removed soil. Start by pushing all the soil to one end of the pan. Go through the soil bit by bit, moving soil to the other side of the pan as you count.

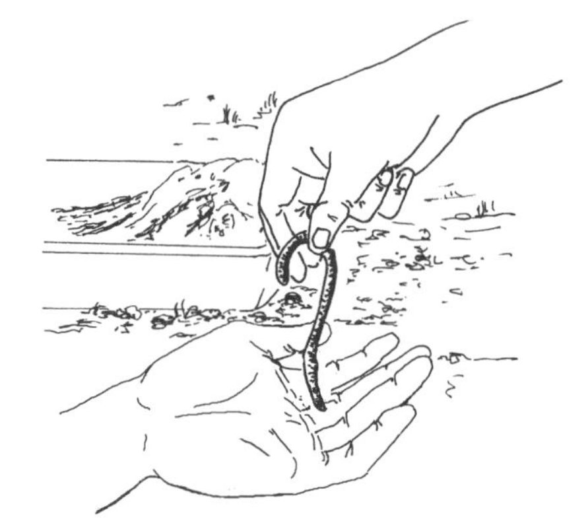

3 If you find only one or two earthworms, your soil needs a lot of help (organic matter.) Five to nine earthworms means you're getting there, but still need more organic matter. If you find 10 or more earthworms, you have healthy, biologically active soil — congratulations!

DID YOU KNOW?

Mary Appelhof, author of the 1982 classic *Worms Eat My Garbage,* must have fantastic soil. When she did a census of the top 7 inches of a square foot of her garden, she came up with 62 worms of all sizes! That's equivalent to 2.7 million worms per acre!

CHAPTER 2

Creating Fertile Soil

In This Chapter

- Nutrient Supply and Demand
- Testing Your Soil
- Meet the Nutrients
- Identifying Deficiencies
- Soil Acidity
- Organic Matter: The Key to Balancing Nutrients
- Forms of Organic Matter

For soil to be ideal, it's not enough for the physical properties (described in chapter 1) to be good. The interlocking chemical properties that create fertility also must be in good shape. Rich loam is created by physical and chemical properties working together.

Fertility means more than just pouring on fertilizers. Plants need a varied and balanced diet, but nutrients are of no use unless they're within reach of each plant's roots (or leaves).

The Great Balancing Act

If nutrients aren't in the right balance, plants can't use them even if they *are* all within easy reach. Too little of one nutrient can partially starve plants by limiting their use of other nutrients. Too much of one nutrient can lock others out of plants' reach. Soil that's very acidic or alkaline locks up several nutrients at once. The best loam is not too acid, not too alkaline, and has a good balance of all essential nutrients.

The largest part of a plant's "diet" is the major nutrients, those used in large quantities (nitrogen, phosphorus, potassium, calcium, magnesium, and sulfur). Plants also need micronutrients (iron, manganese, copper, zinc, boron, and molybdenum), but in minute amounts. Overdoses of any of the micronutrients can be toxic.

Organic matter is your ally in the great balancing act. It provides many if not all the minor nutrients plants need, often in just the right balance, plus small amounts of all the major nutrients. It releases nutrients slowly, matching plant needs and helping to prevent an overdose. Finally, organic matter ensures availability by helping to maintain the right acidity balance.

Nutrient Supply and Demand

Nutrient availability is a constant give and take. In addition to sources supplying nutrients and uses removing them, there are several interacting factors that affect supply and demand. Each is explained in the chart on the next page.

How Nutrients Become Available

- Most nutrients come from the rocks that weathered to become parent material and eventually mineral particles.
- Organic matter supplies many of the same nutrients plus nitrogen, which rarely occurs in mineral form. Organic matter is the remains of animals and plants, which contain many different chemical elements in their cells. As microorganisms break down organic matter, these elements are returned to the soil.
- Nutrients can also be applied to the soil as fertilizer, limestone and other rock dusts, or compost.
- Nutrients have to dissolve in soil water before plant roots can soak them up. The water clinging to soil particles and partially filling pores dissolves nutrients and transports them through soil.

DID YOU KNOW?

Nutrient Reserves

Once nutrients are broken down and made available, they may be stored in the soil's nutrient reserves. Humus and clay are responsible for almost all of the nutrient storage in any soil. As a result, soil with high levels of clay and humus has a tremendous ability to store nutrients for future use.

How Nutrients Become Unavailable

- They may be temporarily withdrawn by plant roots for use in building new leaves and shoots. When plants die and decompose, those nutrients are returned to the soil. Nutrient depletion occurs only if plants are harvested and completely removed from a site.
- Soil organisms and microbes also remove nutrients, but, as with plants, this is temporary. The elements in their cells are eventually returned to the soil when they die.
- Chemical reactions in the soil can cause nutrients to form compounds that don't easily dissolve. This essentially locks up nutrients out of plants' reach.

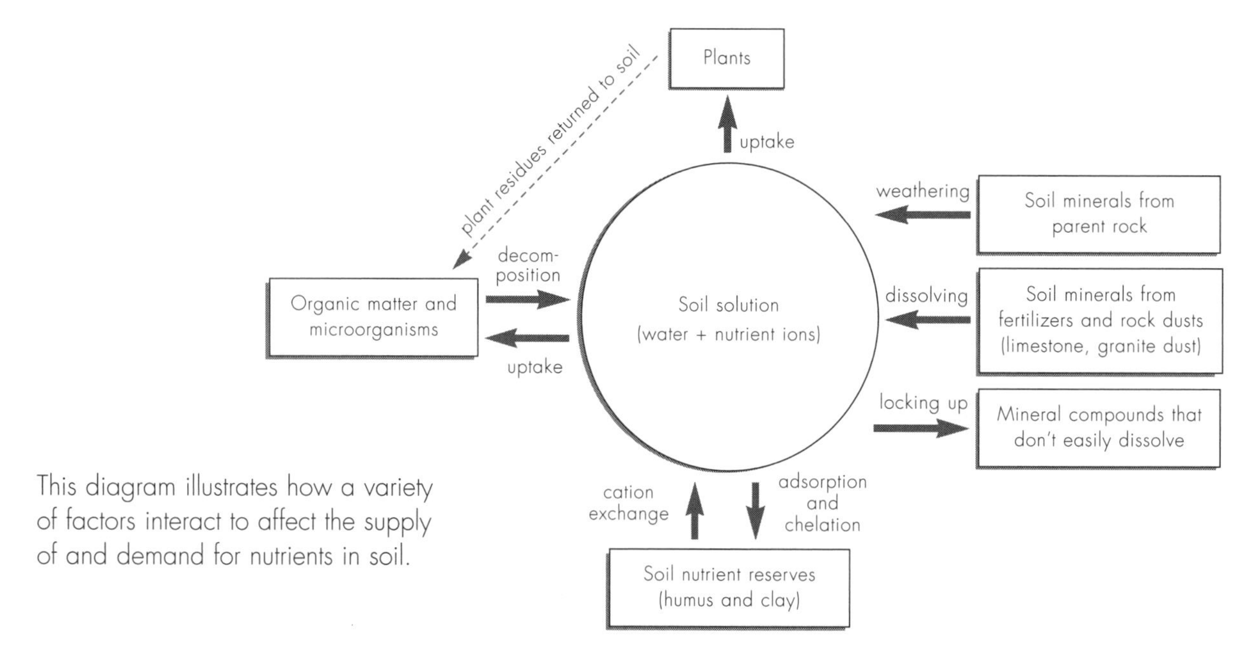

This diagram illustrates how a variety of factors interact to affect the supply of and demand for nutrients in soil.

What Controls Nutrient Supply?	
Soil moisture	In order for nutrients to dissolve, water must be present in the soil. Too little water makes it hard for nutrients to dissolve and move through soil channels. Too much water washes (leaches) many nutrients out of the soil entirely, transporting them to streams or lakes (eventually to the ocean), where they're no use to plants.
Soil air	Soils need good structure — abundant pores and channels — to carry both water and air. In addition to harming plants (roots need air to survive) and soil organisms, lack of oxygen (in waterlogged, compacted, or heavy soils) changes soil chemistry. Lack of oxygen promotes the locking up of some nutrients; it also robs soil of nitrogen by promoting conversion of nitrates into nitrogen gas, which escapes into the atmosphere.
Soil pH	The pH of soil has an even bigger effect on soil chemistry. Either acidity or alkalinity can bind up many nutrients at once. Improper pH is the most common reason for nutrients present in soil to be locked out of the soil solution (and therefore out of reach of roots).
Soil life (microbes)	Microorganisms are largely responsible for breaking down organic matter to release the nutrients it contains. During this process, microbes produce organic acids that dissolve rock and mineral particles. They therefore change soil chemistry on a very small scale, speeding up the weathering process. When soil conditions are good, the organic acids they produce can even dissolve some locked-up nutrients.
Temperature	Since temperature affects microbial activity, it also affects nutrient availability. Warm temperatures increase activity, speeding up decomposition. Cold temperatures greatly slow down activity, and therefore slow the release of nutrients from organic matter. Plants can experience a temporary deficiency of phosphorus or nitrogen in early spring because microbial decomposition in the cold soil is so slow that it can't meet the plants' demands for these two nutrients.
Nutrient interactions	Nutrients may interact with each other to interfere with plant uptake. Too much calcium from overliming can cause an artificial potassium deficiency, for instance, and too much potassium keeps plants from using calcium. It's the balance of nutrients — not just the nutrients themselves — that's responsible for fertility.

Testing Your Soil

Easy-to-use kits are available from garden centers and catalogs for testing your own soil. The simplest measures only pH, but most also measure nitrogen, phosphorus, and potassium. The least-expensive kits are also the least accurate, but if you follow directions with a more sophisticated kit, your results should be fairly good.

While these kits don't provide as many details as a professional test, they tell you something about your soil and they're fun to use. You get results instantly, instead of having to wait six weeks or more. You'll get better results if you supplement kits with a more accurate professional test every few years or — best of all — with close observation of how your plants perform.

MATERIALS

- Soil test kit (make sure all testing equipment is clean)
- Distilled water
- Clean trowel
- Small plastic bucket or bowl (clean and dry)
- Sieve with ⅛- or ¼-inch (.3 or .6 cm) mesh (optional)

The first three steps are essentially the same ones used to prepare a sample for a professional test. Some labs require a slightly different method; be sure to follow the specific directions given by the lab.

HINTS FOR SUCCESS

- For the most accurate results (especially where water is very hard or soft), use distilled water for all tests. You don't have to go out and buy it if you know anyone with a dehumidifier — the water dumped out of a dehumidifier is distilled, and it's free. Use distilled water to rinse out test tubes when you're done.
- Do separate tests for various garden areas (lawn, vegetable beds, and flower beds).
- Test while soil is moist, not soggy wet or powdery dry. Cold soil gives low readings for nitrogen and phosphorus.
- Avoid testing unusual spots such as dips and next to a compost pile.

1 Clear away plants and thatch, leaves, and other litter from the soil surface. Dig a 6-inch-deep (15 cm) hole in garden beds; if you're testing lawn areas, a 4-inch (10 cm) hole is deep enough.

Master Gardening Tip

Getting a Professional Soil Test

A professional test is a good idea when you start a new garden. Testing periodically (every four to five years) after that lets you know if you're on the right track.

You can have your soil tested by either a public (state) or a private laboratory. To locate state labs, call your local Cooperative Extension Service for information and prices. (The Cooperative Extension Service may be listed under county, state, or federal government in the phone book.)

Public labs. Most state labs want you to use their kits for collecting samples. Others have specific requirements for how to collect samples; be sure to follow these for accurate test results. You'll get the best results from your own state, as recommendations will take local soil types into account. Be sure to tell the lab how you intend to use the soil being tested (lawn, vegetable, or flower garden).

Private labs. If you want a more detailed soil analysis than that provided by most state labs, try a private lab. (A list is provided on page 205.) Be prepared to pay more. In addition to testing, these labs provide outstanding, thorough, customized reports. Some offer hybrid recommendations using both organic and synthetic fertilizers for soil improvement; most will provide strictly organic recommendations if you request them. A few offer free consultation after the test.

2 Carve a thin, 6-inch-long (15 cm) slice of soil from the side of the hole and place it in the bucket. Dig four to six more holes evenly spaced over the test area. Remove a slice from the side of each and place in the bucket to create a single, average sample.

3 Mix the samples well and break up as many crumbs as possible. Remove any large bits of plants or pebbles. You may find it easier to sift the mixed soil to remove larger particles and ensure that the remaining material will be small enough to fit inside the kit's test tubes.

4 Follow kit directions carefully for the amount of soil and amount of indicator chemical to use for each test. Mix as directed, allow to settle for the specified amount of time, and compare the test tube to the colored chart supplied in the kit.

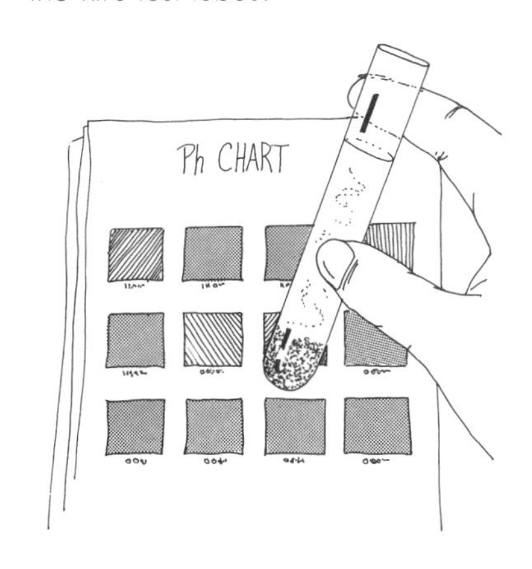

Meet the Nutrients

What are the nutrients that plants require for good health, and why do plants need them? Most are common chemical elements. They fulfill one or more of three general roles. Some are structural — part of plant cells, and therefore their basic architecture. Others are metabolic, required for any of the many biochemical functions within plants, from photosynthesis to reproduction. The third group are catalysts; these are substances that assist metabolic functions without being used up in the process. A list of nutrients needed by plants, along with a phrase to help you remember them, is given in the box at right.

Nutrients for Free — Carbon, Hydrogen, and Oxygen

Three elements that plants need — carbon, hydrogen, and oxygen — never appear on any fertilizer label, even though they make up almost 95 percent of any plant. That's because plants can easily get all they need from the air and from water. Plants use lots of each, but gardeners get a free ride! (You can starve a plant of oxygen if you grow it in waterlogged soil, though; see "Improving Soggy or Poorly Drained Soils" on page 138.)

Major Nutrients

After the freebies, the elements plants need in the largest quantities are called major nutrients. These are the elements most likely to limit plant growth in gardens. That's why they're the ones you're most apt to recognize from fertilizer labels. The major nutrients are nitrogen, phosphorus, potassium, calcium, magnesium, and sulfur. They're sometimes called macronutrients.

Three of these major nutrients are most familiar to gardeners because they're the ones that appear in standard fertilizers: nitrogen, phosphorus, and potassium. Most soils that are gardened or farmed intensively need periodic resupplies of all three. Often — but not always! — they must be applied every growing season, which is why they're included in standard fertilizers.

Three other major nutrients — calcium, magnesium, and sulfur — already exist at adequate levels in some soils. They may not need to be applied at all, or at least not as often as the previous three.

Gardeners with acidic soils who apply lime every few years know the importance of calcium (and sometimes magnesium) to soil chemistry. They may not need to add sulfur, though. Gardeners with alkaline soils usually face just the opposite situation. They may be used to adding sulfur periodically to adjust soil chemistry, countering the effects of too much calcium and/or magnesium.

HINT FOR SUCCESS

How to Remember the Essential Elements

Agriculture students use a phrase to help them remember the elements needed for plant growth and development. The phrase is made up of the chemical symbol for each element, as listed below.

C HOPKNS CaFe Mg B Mn CuZn Mo Cl NiCo is short for "See Hopkins' café managed by mein cousins Moe and Clyde Nico."

The nutrients included in the mnemonic phrase are:

C carbon
H hydrogen
O oxygen
P phosphorus
K potassium
N nitrogen
S sulfur
Ca calcium
Fe iron
Mg magnesium
B boron
Mn manganese
Cu copper
Zn zinc
Mo molybdenum
Cl chlorine
Ni nickel
Co cobalt

Micronutrients

Micronutrients are also called trace elements because only a very small amount — a trace — of each is required for healthy growth. They include iron, boron, manganese, copper, zinc, and molybdenum.

The importance of these nutrients has become more obvious with increased use of synthetic fertilizers. When farmers and gardeners had to depend on manures for major nutrients, they didn't have to worry about micronutrients. The current generation of synthetic fertilizers is so purified that few if any micronutrients remain in them. As a result, deficiency symptoms (once thought to be diseases) are now more common. Crops grown on soils lacking these nutrients may fail to supply human dietary needs, even if plants don't appear to suffer.

You may never have to worry about micronutrients if you routinely add compost or manure, or if you use fish emulsion and/or liquid seaweed as fertilizers from time to time. Micronutrients should never be added unless soils are known to be deficient, and then only in the recommended amounts. The difference between helpful and harmful amounts is incredibly small. Also, deficiencies of micronutrients are often irregular, occurring only in some areas of a garden. Pockets of poor drainage, where soil pH differs from the surroundings, or where soil has washed away are likely spots for small-scale deficiencies.

DID YOU KNOW?

How Small Is a Trace?

To give you an idea of the difference between major and trace nutrients, consider the needs of corn. During the growing season, an acre of corn can easily withdraw 150 pounds (68,000 g) of nitrogen from the soil. But those same plants might withdraw only half an ounce (15 g) of the trace element boron from an acre. That means corn needs over 4,500 times as much nitrogen as boron in its diet!

Chlorine, Nickel, and Cobalt

Like carbon, oxygen, and hydrogen, chlorine and nickel don't need to be applied to soils. Researchers have proved that plants can't grow without them, but gardeners don't need to worry about them.

Cobalt is required by some plants and many microorganisms, including nitrogen-fixing bacteria. The amount of cobalt needed is too small even to measure; so again, don't worry about it.

Beneficial but Not Essential

The element silicon doesn't qualify as essential because many plants grow well without it. But some plants benefit from added silicon, so think of it as a beneficial rather than an essential nutrient. Corn contains high levels of silicon, which helps make cornstalks stiff. As you might expect, corn is one of the plants that benefits from added silicon. Silicon also appears to increase disease resistance. Sand (silica) is naturally rich in silicon. Greensand (a special type of sand that's green in color) is best for adding silicon to soil; it's also a good source of potassium. Granite dust is another good source. Both of these are also excellent slow-release sources of essential micronutrients.

Identifying Deficiencies

A deficiency of an essential nutrient harms plants by slowing or stopping particular functions. In order to grow, plants have to produce new cells. If a nutrient required for cell division is lacking, the plant can't make enough new cells, so growth slows dramatically or even stops.

Looking for Signs

By the time nutrient deficiencies become visible, some damage has already been done. A soil test can detect deficiencies before they cause damage. However, since soil changes over time, and since you can't test the soil in every part of your garden, it helps to know how to recognize common nutrient deficiencies. These are easiest to recognize when they're severe and hardest to recognize when more than one is present at one time. Consult the chart on pages 204–205 to help match your plant's symptoms to a probable cause. Try to figure out the cause of any nutrient imbalance before you attempt to correct it. If the "deficiency" is caused by a physical soil problem (see box at right) or interference from too much of another nutrient (which will show up in a soil test), you won't fix anything by dumping on fertilizer. Once you identify a specific deficiency, turn to chapter 6 for information about the best sources of that nutrient.

Is It Really a Deficiency?

Before you assume you're missing nutrients, rule out other causes. Addressing such causes can alleviate deficiency symptoms with no added fertilizer. Using a foliar spray of a liquid seaweed/fish emulsion combination will keep plants growing until long-term improvements occur.

- **Environmental problems.** Drought, heat, and cold (or growing a variety or plant not suited to your climate or soil) causes poor yields more often than does a lack of fertilizer. Low temperatures (as in early spring), overwatering, and underwatering can make nutrients temporarily unavailable.

- **Pests or diseases.** Some pests and plant diseases can cause many of the same symptoms as nutrient deficiencies. If only one type of plant is doing poorly, look for a critter as the culprit. Root-knot nematodes can impair plants' ability to absorb existing nutrients. Aphids and spider mites cause yellowed leaves. Viral diseases can cause mottled, yellow, or deformed leaves.

- **Physical soil problems.** A nutrient may be present but locked up by poor aeration or poor drainage. Alkaline or very acidic soil is the most common cause of locking up available nutrients. To maximize nutrient availability, try to maintain a pH of 6.3 to 6.8.

DID YOU KNOW?

Disease or Deficiency?

Several plant "diseases" besides chlorosis (iron deficiency) are now known to be caused by a nutrient deficiency:

- ***Blossom-end rot*** affects tomatoes, melons, and bell peppers. It's caused by too little calcium, often appearing when plants are stressed by lack of water.
- ***Withertip*** is merely a lack of copper.
- ***Gray speck*** of oats, ***speckled yellows*** of sugar beets, and ***marsh spot*** of peas are all caused by insufficient manganese.
- ***Hollow heart*** of cauliflower, beets, and turnips is caused by too little boron.
- ***Corky areas*** in apples are also from boron deficiency.
- ***Yellow spot,*** which afflicts citrus fruits, is merely a lack of molybdenum.
- ***Whiptail,*** which is recognized by weirdly distorted, long leaves of cabbage and related plants, is also caused by molybdenum deficiency.

Symptoms on upper leaves? Some nutrients stay put; they can't move within the plant because they're built into the cell structure. Sulfur, calcium, iron, manganese, boron, and copper are needed to make healthy new leaves, but plants can't borrow them from elsewhere if there's not enough. As a result, deficiency symptoms of these structural elements show up first on young, upper or outer leaves.

Symptoms on lower leaves? Some nutrients get around — they're easily shuttled around within the plant: nitrogen, phosphorus, potassium, magnesium, molybdenum, and zinc. That means if plants aren't getting enough, they can move these to young, growing tissue, where they're needed most. As a result, deficiency symptoms of these mobile elements show up first on older, lower leaves, eventually spreading to the rest of the plant.

Soil Acidity

The acidity of soil is one of the most important factors behind fertility. It affects nearly all soil properties, from chemistry to biological activity to structure. Correcting soil acidity or alkalinity may improve your garden more dramatically than any other soil-building effort.

Acidity is caused by the amount of hydrogen ions present in soil. When you test for your soil's acidity or alkalinity, what you're really measuring is the concentration of hydrogen ions.

Acidity Affects How Nutrients Dissolve

Acidity controls nutrient supply by controlling how well nutrients dissolve. In very acidic and alkaline soils, most nutrients dissolve very slowly or not at all. They may form insoluble mineral compounds, locking up nutrients so that plants can't use them. If your soil's too acidic or alkaline for nutrients to dissolve easily, you can waste a lot of money on fertilizer. Plants will be able to reach very little of what you apply. That's why it's important to test and correct the pH of your soil *before* you add fertilizer. (For ways to test pH, see the facing page; to correct pH, see chapter 7.) By correcting extreme acidity or alkalinity, you can bring more nutrients within plants' reach without adding any fertilizer! This is especially true for nitrogen, phosphorus, potassium, sulfur, molybdenum, and boron.

DID YOU KNOW?

Sweet and Sour Soils

- ***Acidic soils*** are sometimes called sour soils because substances that are acidic often taste sour. Vinegar is a common acid that tastes sour; its pH is between 3 and 4. (Don't try tasting strong acids such as sulfuric acid, though — they'll burn a hole through your tongue!) Farmers used to test their soil by tasting; they could tell it was too acidic if it tasted sour.
- ***Alkaline soils*** used to be called sweet soils since they're the opposite of acidic soils. They don't taste sweet, though. Any old-time farmer who learned to recognize alkaline soils by tasting will tell you they're bitter.

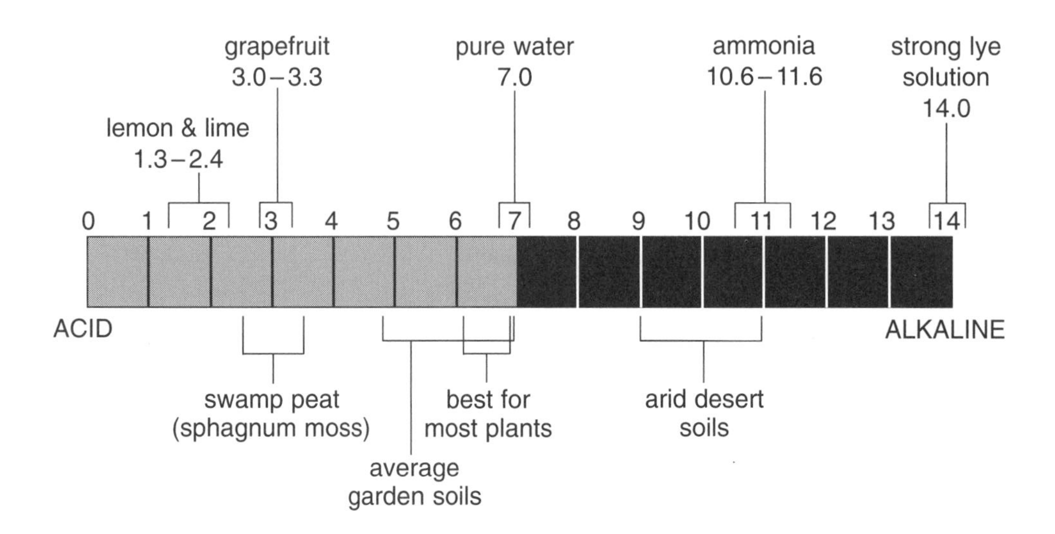

The scale for measuring pH runs from 0 to 14. The pH scale above shows the acidity or alkalinity of some common substances. Neutral soil has a pH of 7. Lower numbers mean increasing acidity; higher numbers mean increasing alkalinity. The pH at which each nutrient is most available varies slightly. The ideal range to maximize nutrient availability (and therefore plant health) is 6.3 to 6.8. If soil contains ample organic matter, many plants will tolerate a pH of between 5.5 and 8.0.

Testing Soil Acidity

Litmus paper is an inexpensive way to test your soil's pH. It comes as either short strips or a roll of ¼-inch-wide (.6 cm) paper saturated with special dyes. Touch a short piece to garden soil mixed with a little distilled water. The paper will change color. Compare this color to the chart that comes with the litmus paper to find the pH value for your soil. It's accurate to within about half a pH unit.

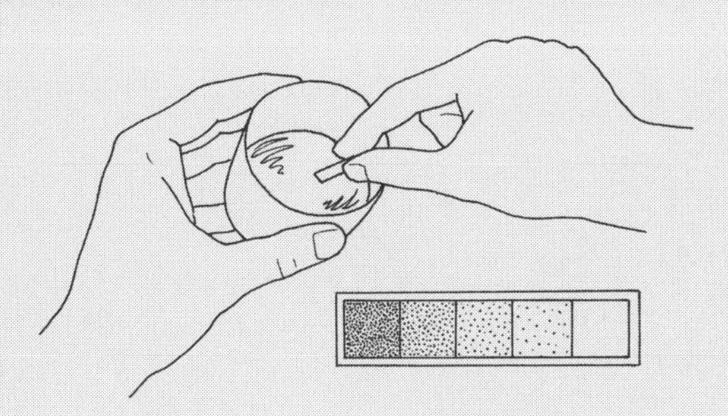

A pH meter is a more expensive, but more accurate, option. It's a simpler, portable version of the equipment used by soil laboratories to test pH. Both use electrodes to measure hydrogen ion activity. Similar meters are available for testing soil salt levels.

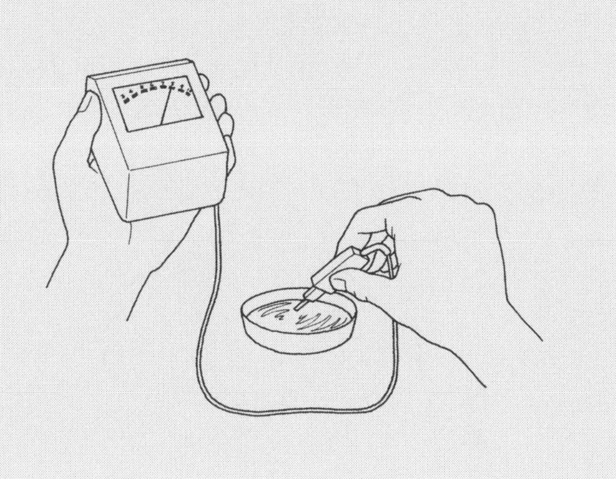

Soil pH in the United States

Acidic soils are common where rainfall is abundant. Rain washes out the chemicals that counteract acidity. Gardeners east of the Mississippi River and in the Pacific Northwest generally have acidic soils and need to raise soil pH. Alkaline soils are common in areas with little rainfall. Gardeners in the southwestern United States and much of the West generally have alkaline soils and need to lower pH.

Organic Matter: The Key to Balancing Nutrients

Organic matter is the magical stuff that transforms a pile of sand or fine rock particles — or a lump of potter's clay — into fertile soil. It's just plants in various stages of decomposition. Think of what you see on the ground in a forest: fallen leaves and twigs, bits and pieces of leaves, and — just above the soil — a layer of dark brown leaf mold. All this is organic matter. Even manure is really just processed plants; pieces of straw are usually still visible in fresh manure.

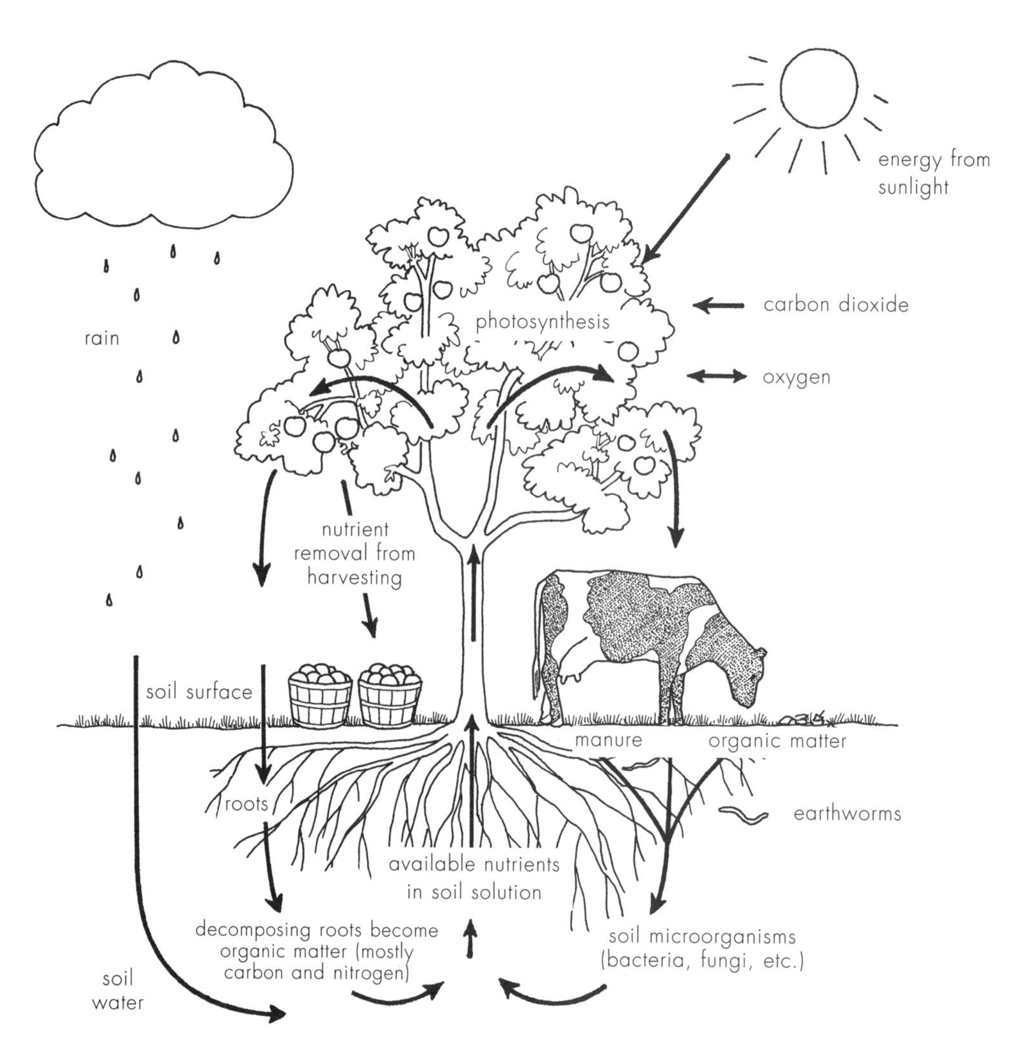

Natural ecosystems have a self-sustaining recycling program for organic matter. Plant leaves harness the sun's energy to manufacture carbon-containing energy compounds (carbohydrates) from carbon dioxide in the atmosphere. Plants use these energy compounds, plus nutrients absorbed from the soil, to manufacture proteins and all the other compounds they need for healthy growth. When leaves fall to the ground, they become the food of earthworms and other organisms. Soil organisms convert these plant parts, plus the wastes of animals that eat the plants, into humus and nutrients. Roots absorb these nutrients for plants to use in growing new leaves. New leaves create new sources of organic matter, and the cycle begins all over again.

Types of Organic Matter

Organic matter includes fresh plant residues, as when crops are turned under at the end of the growing season. Roots and stems left behind after harvest, or the leaves on a forest floor, often make up a significant proportion. Partially decomposed plant residues such as those in rough compost are a particularly useful form for gardens. The final stage of organic matter, when microorganisms have processed plants until they're completely unrecognizable, is called humus.

Humus is a fine, dark substance that only looks relatively uniform. Chemically, it's varied and complex. It's much more resistant to breaking down than earlier stages of organic matter. Some forms last many years. Humus gives rich, fertile soil its dark color. It promotes healthy plant growth in many ways, from keeping existing soil nutrients available to enhancing soil structure for good root growth.

What Organic Matter Can Do

Organic matter is essential for maintaining a steady supply of nutrients, especially nitrogen and potassium. Fresh plant residues and manures supply all nutrients. Well-decomposed forms and humus supply micronutrients and low levels of nitrogen, phosphorus, and sulfur but little calcium, magnesium, or potassium.

Any form of organic matter helps keep nutrients from being washed away by heavy rains, yet miraculously holds these nutrients loosely so they're easily released into the soil solution to meet plant needs. Finally, humus naturally buffers soil acidity, helping to keep it just slightly acidic; at this pH, the availability of existing soil nutrients to plants is greatest.

Why You Need to Add Organic Matter

Gardening speeds up — and simultaneously removes plant material from — the cycle shown at left. It helps microbes work more efficiently by supplying air (through digging soil) and a steady supply of moisture (through watering crops). That efficiency is good when your soil is well stocked. But harvesting fruits and vegetables, mowing the lawn and removing clippings, raking leaves, and cleaning out beds at the end of the season take organic matter out of circulation. If you don't add organic matter, its level — and your soil quality — will drop quickly.

DID YOU KNOW?

How Organic Matter and Humus Work for You

- Improves soil's ability to absorb rainfall or irrigation water while reducing surface runoff.
- Improves soil's moisture-holding ability.
- Provides food for beneficial organisms such as earthworms.
- Helps suppress harmful soil organisms such as nematodes and some diseases.
- Buffers soil pH while increasing plants' ability to tolerate acidic or alkaline soils.
- Supplies nutrients in ideal slow-release form.
- Supplies micronutrients lacking from most commercial fertilizers.
- Helps prevent vital nutrients from washing away.
- Reduces fertilizer requirements.
- Is transformed into vitamins, growth hormones, and other substances that stimulate plant growth and soil organisms.
- Aerates soil, increasing pore space for good air and water movement.
- Improves structure and drainage of all soils, but especially those high in clay.
- Improves sandy soils by increasing their retention of water and nutrients.
- Reduces soil erosion.
- Helps keep soil cooler in summer, warmer in winter.
- Helps inactivate toxins such as pesticides and heavy metals in contaminated soils.

Forms of Organic Matter

Now that you understand how important organic matter is, how do you get it into your garden? There's a wide choice of readily available sources, and several easy techniques you can mix and match according to your interests and energy level. Here's an overview; for more details, turn to chapters 4 and 5.

DID YOU KNOW?

Compost Reduces Pests and Diseases

Compost is more effective than other forms of organic matter for controlling pests and diseases. Tests show the problematic azalea lacebug is less apt to bother azaleas growing in compost-enriched soil than azaleas growing in soil enriched with peat moss.

Compost has fungicidal effects, also. Large-scale greenhouse and ornamental plant operations have reported reduced needs for soil fungicides for problematic diseases such as damping-off and fusarium when compost is used as a major ingredient in their potting soils. Many gardeners have had success controlling leaf diseases by soaking compost in water and spraying this "compost tea" directly onto the diseased leaves. (See page 129 for how to make compost tea.)

Compost

Compost has the advantage of being easy to make yourself. Now that hauling leaves and yard trimmings to the dump is no longer an option for many people, compost is becoming recognized as an important way to "dispose" of such materials. A more positive view of this "disposal" is that compost offers a way to make a high-quality soil amendment and low-level fertilizer out of free materials. For people who prefer not to make their own, commercial compost can be ordered by the truckload or bought in bags at garden centers. Compost is the best choice for increasing pest and disease resistance.

Plant Residues and By-Products

Many plant residues supply low levels of major nutrients as well as many micronutrients. These range from lawn and garden trimmings to by-products from processing plants. Where readily available, they're often the least-expensive option. Sometimes they're free. Think of plant by-products as the raw materials of compost; you have to wait for them to break down to supply all their associated benefits.

If you live right on the ocean, you may be able to gather seaweed. It's an especially rich source of potassium and micronutrients as well as a good form of organic matter. If you have lots of deciduous trees in your yard, you're already good at collecting leaves. To encourage rapid decomposition, leaves should be chopped. You can easily chop leaves by running a lawn mower over them (or by running them through a special leaf shredder).

Contact vegetable- or fruit-processing plants in your area. If you live near a cider mill, ask about apple pomace, the residue left after apples are pressed into cider. Wineries may be happy to supply grape pomace. Sawmills produce lots of sawdust; furniture makers may produce sawdust and/or wood shavings.

Sawdust, wood shavings, and straw are lightweight and easy to handle. All are much higher in carbon than are other sources of organic matter. As a result, they temporarily tie up nitrogen from the soil as they break down. If you use only a thin layer, this won't be a problem. If you want to apply more than an inch (2.5 cm) at a time, you'll need to supply extra nitrogen (as by mixing with grass clippings or dehydrated manure). Or you can dig abundant amounts into the soil in fall for spring planting.

Animal Manures

Animal manures have been an important source of organic matter for centuries. They offer all the benefits of other sources, with just a couple of disadvantages. Fresh manures are much richer sources of nitrogen than most organic matter, but they may burn plants. Aged manures offer lower levels of nutrients, but there's no risk of burning. Horse and cow manures may contain seeds and therefore introduce weeds into your garden. Many people like to use them anyway; they just compost them or use thicker mulch to keep weed seeds from sprouting.

For people unfamiliar with farming, another warning is in order: Chicken manure may contain a fungus that can cause a nasty respiratory infection in humans, one that's difficult to cure. To be safe, wear a respirator when handling chicken manure unless it has been composted.

In more rural areas, manures are readily available and can be delivered by the truckload. In any area, dehydrated and composted manure can be purchased in bags from garden supply centers.

Green Manures

One form of organic matter presents a special case. Green manure is both a type of material and a method. Green manures let you grow your own organic matter right where it's needed. All you spread is seeds; you wait for the plants to grow before doing any turning under. While you need to watch the timing of turning the crop under to maximize the benefits, green manures are otherwise remarkably easy to use.

HINTS FOR SUCCESS

Three Ways to Apply Organic Matter

In addition to choosing one or more sources of organic matter, you need to decide how to add it to the soil. One or more of the following three methods will work for most sources.

- ***Soil amendment.*** Mix the material into soil.
- ***Topdressing.*** Spread material on top of soil (ideally it should disappear fairly quickly so it's available to plants).
- ***Mulch.*** Really just a year-round, slow-acting topdressing (ideally it should stay around longer so you don't have to replenish it frequently).

How Much Does Organic Matter Weigh?

Although the weights here are approximate, they'll give you a pretty good idea of how heavy these materials are if you're hauling them around. They'll also help you figure out what volume you'll need to work with if you know the weight required. A bale of hay (14 inches x 20 inches x 32 inches) weighs about 40 pounds. *(For metric equivalents, see "Useful Conversions" on page 208.)*

Material	Weight of 1 bushel	Weight of 1 cubic yard
Compost	50 lbs.	1,000 lbs.
Horse manure	40 lbs.	800 lbs.
Poultry manure	50 lbs.	1,000 lbs.
Other manures	80 lbs.	1,700 lbs.
Peat moss (compressed)	25 lbs.	500 lbs.
Sawdust	15 lbs.	300 lbs.

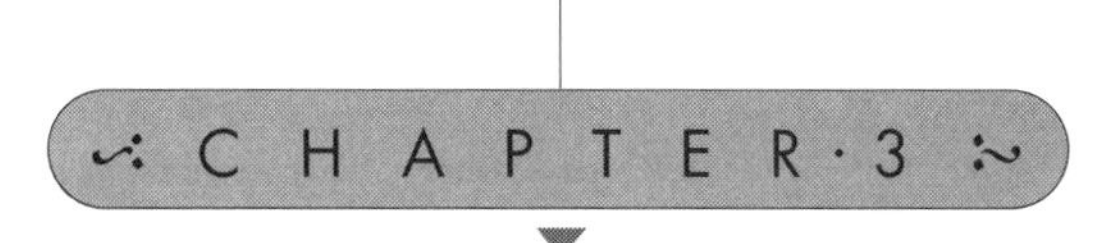

Soil-Building Tools

In This Chapter

- Basic Hand Tools
- Long-Handled Tools
- Types of Hoes
- Sharpening Your Tools
- Maintaining Hand Tools
- Protecting Your Back
- Power Tools
- Rototillers and Cultivators
- Maintaining Power Tools
- Storing Tools
- Handy Haulers

Loosely, *cultivation* refers to any form of working the soil or general gardening. The word is used in a more specific sense to mean breaking up or loosening soil. This includes preparing soil for planting and even weeding. In this book, and especially in this chapter, cultivation refers to manipulating the soil to loosen it.

Hand tools for cultivating and weeding haven't changed a lot for several hundred years. Materials and manufacturing methods have been refined, but most of today's tools would be recognized by someone from the Middle Ages.

Get By with the Basics

While it's fun to drool over fancy tools in lush catalogs, you don't need very many to get by. In general, try to buy one really good hand tool rather than two inexpensive ones. Well-made tools are a worthwhile investment because they're less apt to break. In addition to working better and lasting longer, they feel better in your hand.

Power tools reduce the effort needed for strenuous tasks. Gardeners managing a half acre (hectare) or more may not want to do without them, but they're not essential. A wheel hoe requires little effort to accomplish the same tasks. In very small gardens, using hand tools is often less of a hassle than firing up gas-powered engines. For small, level lawns a non-powered reel mower is easy, quiet, and needs no maintenance except periodic sharpening. Electric tools are a good compromise for providing power in smaller gardens in exchange for minimal maintenance. In medium-size gardens, you may find tillers or cultivators and chippers useful, but you may need them only rarely. Consider renting such tools once a year, to see which you prefer and how often you really need them.

Basic Hand Tools

Hand tools are essential for many garden jobs. Every gardener needs at least one trowel, and some form of weeding tool is also important. Many variations of weeders are available, so look around for the one that fits your hand and best suits your needs. If you're on a limited budget, ask to try out friends' weeding tools before buying your own.

Different tools come in handy for different tasks. Use a trowel to dig holes for small transplants or bulbs. Hand cultivators are great for mixing in phosphorus or limestone around perennials and for digging up weeds, roots and all. To remove a large patch of chickweed, though, a sharp hand hoe is much faster. You can slice off chickweed just below the surface with only a couple of swipes. Wide hand hoes cover larger areas quickly; small or narrow blades let you get between closely spaced plants. The one hand tool that comes closest to being all-purpose is the Asian hand cultivator — it can dig small holes, scratch soil, and remove weeds.

Before You Buy

When shopping for hand tools, pick each one up to make sure it feels comfortable. Single-piece, cast-aluminum construction is very durable. Avoid single-piece, pressed-metal tools, as they bend and break much more easily.

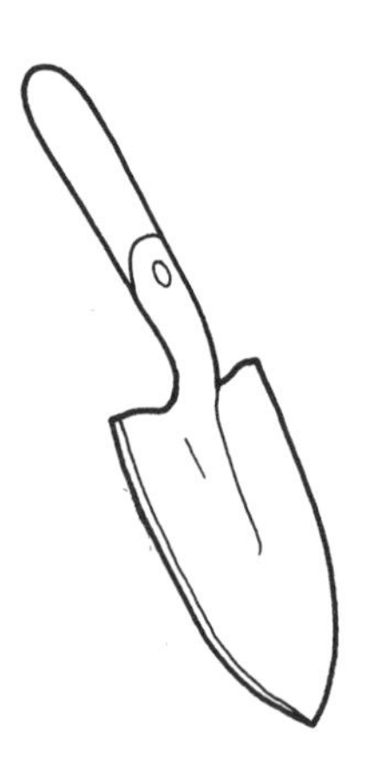

Wooden-handled tools are very durable if well made. Their blades should be forged steel rather than pressed metal. Look for a blade whose socket wraps all the way around the handle and is secured by a pin through the shaft. Or look for strips of metal that are bolted to the handle. Avoid tools whose blades are simply pressed into the handles.

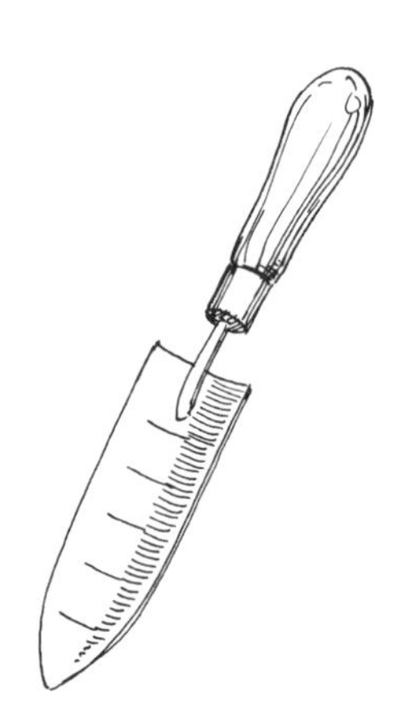

Trowel. Available in various widths and lengths, with shallow to deep scoops; some have inches marked on blades. Used for general small-scale digging, transplanting, planting bulbs.

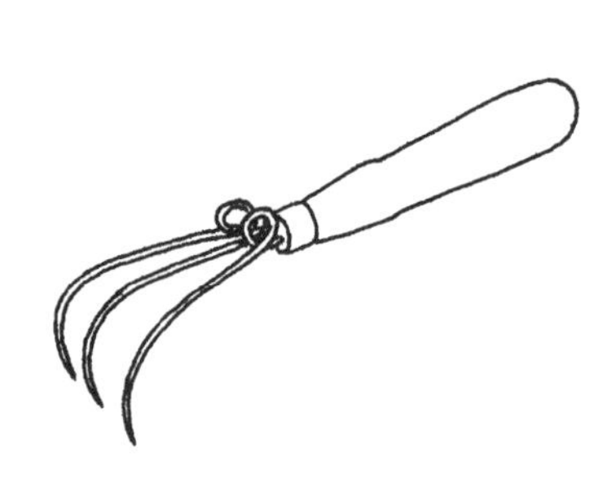

Hand cultivator. Used with a pulling motion. Spring-tined models are less apt to break on rocks. Good for weeding around plants, loosening soil, scratching in fertilizer.

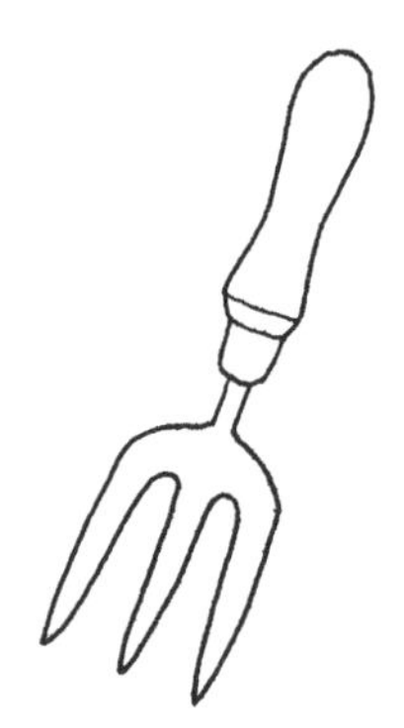

Hand fork. Used with a digging motion (it's awkward to use with a pulling motion). Good for transplanting, weeding, loosening crusted soil, mixing in fertilizer.

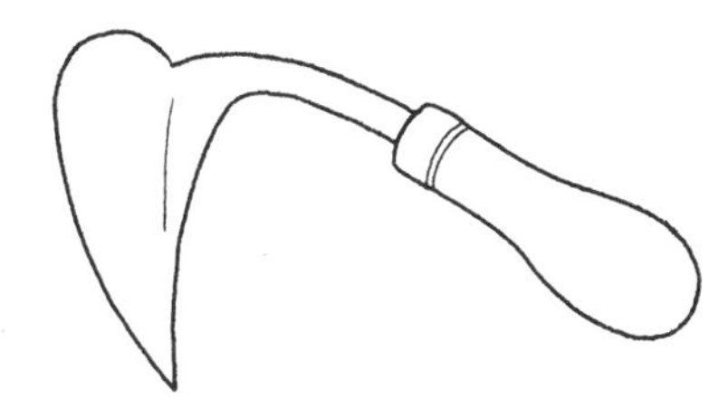

Asian hand cultivator/Ho-Mi. Very versatile, ancient design; used with a pulling motion. Good for general digging and cultivating, weeding, opening and closing furrows, transplanting.

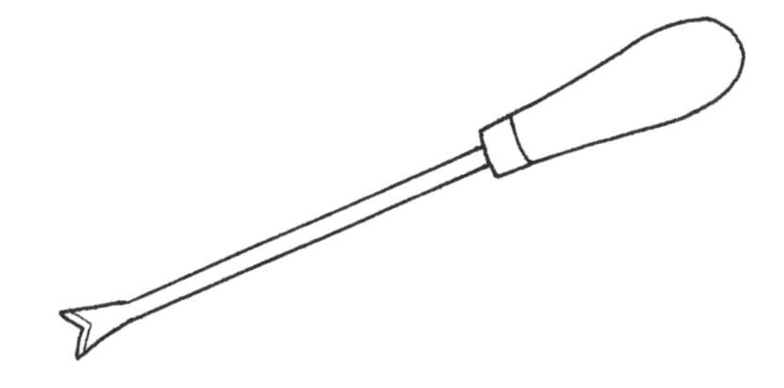

Dandelion weeder/Asparagus knife. Good for digging out weeds with deep taproots or any weeds in narrow spots; also used to harvest asparagus by cutting spears just below soil level.

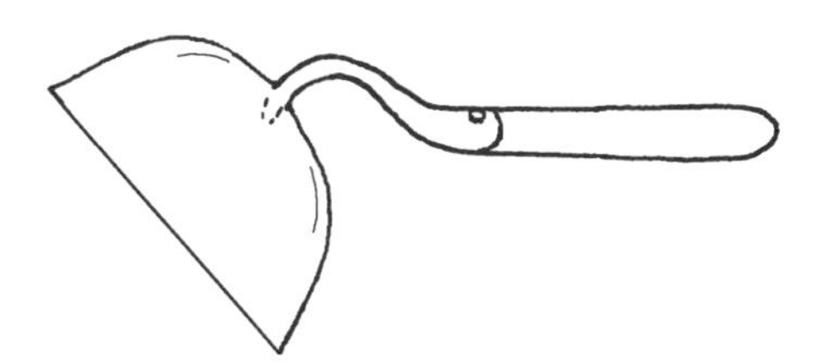

Hand hoe/Dutch weeder. Blade is pulled below the soil surface to slice weeds off their roots. Also used for loosening soil.

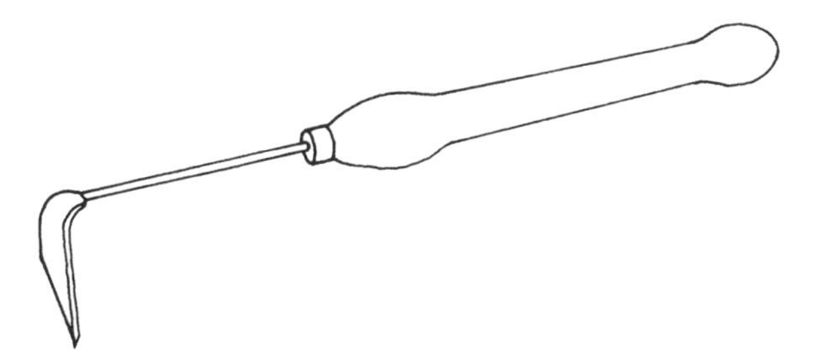

Cape Cod weeder. Used for loosening soil and weeding; blade is pulled below soil surface to slice weeds off at their roots. Pointed end digs out tough, deep-rooted weeds; narrow blade allows weeding in tight spots.

Hoses

Most people don't think of hoses when they think of tools, but they're essential for soil building as well as for watering plants. Invest in a quality hose. Cheap hoses have to be replaced or repaired too often.

If the hose doesn't come with a plastic collar at its faucet end, install a hose saver (a short piece of hose inside a heavy metal spring) or brass gooseneck coupling at the faucet to extend your hose's useful life. If your hose gets cut or springs a leak, it's simple to repair with a kit available at any garden center or hardware store. Be sure to buy one that's the same diameter as your hose.

If you're trying to water a large area, you'll need some form of watering device at the end of your hose. These range from simple soaker hoses, nozzles, and sprinklers to elaborate drip irrigation systems. Although more expensive, drip systems can pay for themselves where water use is restricted or water rates are high.

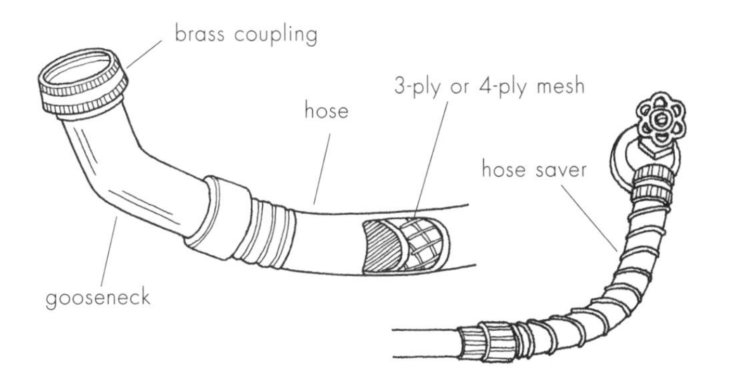

A good hose has a diameter of 5⁄8 to 3⁄4 inch with at least 3-ply, preferably 4-ply, construction. (*Ply* refers to the layers of mesh reinforcing.) It should tolerate at least 500 pounds of pressure per square inch (psi). Cast-brass couplings are best for resisting dents and crushing.

Long-Handled Tools

Choose long-handled tools using the same criteria as for other tools. While you can get by with just a shovel and a garden fork, you'll find garden work easier if you own both a short-handled spade and a long-handled shovel, plus a garden rake for smoothing beds. A sharp, lightweight, cultivating hoe is a worthwhile investment for weeding all but the smallest gardens. Hoes parade under a variety of confusing names, so go by the shape rather than the name. Many more variations are available in addition to those shown here.

Some of these tools come in two styles: with either a straight, shoulder-high handle or a waist-high handle ending in a grip. Shoulder-high handles provide better leverage for shoveling; the shorter handles are easier to maneuver, especially in small spaces. Standard waist-high handles measure 28 inches (70 cm). Tall people should seek out the harder-to-find 32-inch (80 cm) handles. The best handles are made of ash with a straight, tight grain running the length of the handle, or of fiberglass. While good fiberglass is stronger than good wood, poor fiberglass is weaker than wood. Solid-core fiberglass is the most expensive and practically unbreakable. Hickory is heavier and less flexible than ash; choose it only for heavy-duty tools with short or medium-length handles (mattocks and grub hoes).

For the working part of the tool, forged or heat-treated, tempered steel is much stronger (and more expensive) than pressed metal. The lower the gauge of steel, the thicker the metal. With shovels, 16-gauge is fine for anyone except a landscape contractor; avoid weaker 18-gauge shovels. Longer sockets are stronger, particularly if they're slightly longer in the front than in the back.

Before You Buy

Choose the best-quality long-handled tools you can afford, and heft a tool to be sure it feels comfortable before purchasing it.

Master Gardening Tip

Styles of Grips

The grips on waist-high handles come in three forms: D-shaped, split Y-D, and T-shaped. All work well; choose the grip that feels most comfortable.

Shovel. Scooped blade for lifting soil, sand, gravel, manure. A rolled top edge makes the best footrest. Deeper bowls hold more but make lifting harder.

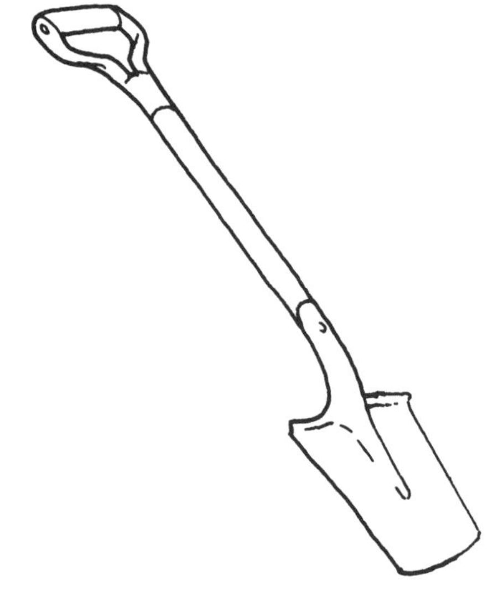

Spade. Flat blade for digging, turning soil; also good for removing sod, edging, dividing perennials. Border spades have slightly smaller blades and are therefore lighter; transplanting or drain spades (also called poacher's spades) have long, narrow blades.

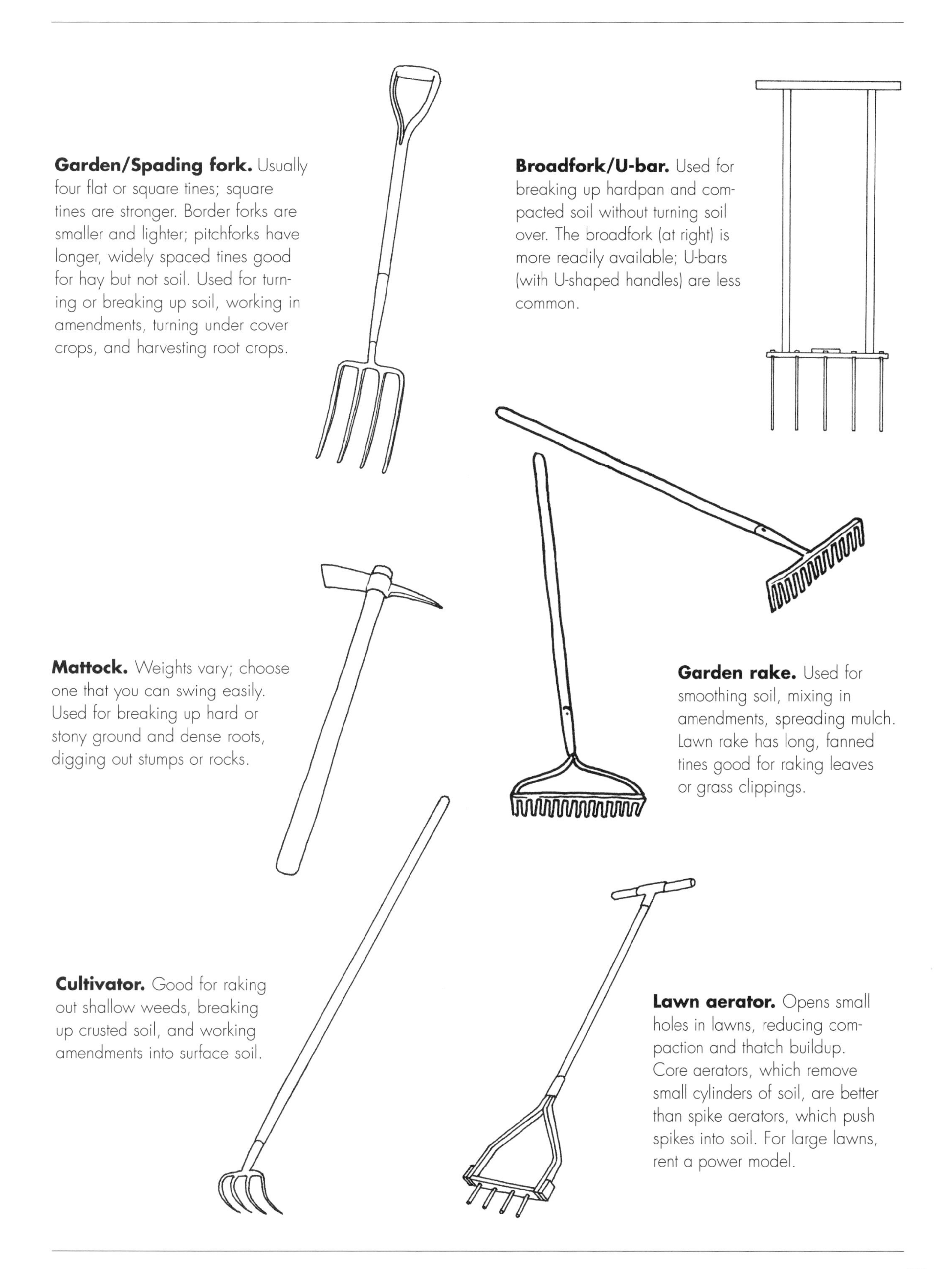

Garden/Spading fork. Usually four flat or square tines; square tines are stronger. Border forks are smaller and lighter; pitchforks have longer, widely spaced tines good for hay but not soil. Used for turning or breaking up soil, working in amendments, turning under cover crops, and harvesting root crops.

Broadfork/U-bar. Used for breaking up hardpan and compacted soil without turning soil over. The broadfork (at right) is more readily available; U-bars (with U-shaped handles) are less common.

Mattock. Weights vary; choose one that you can swing easily. Used for breaking up hard or stony ground and dense roots, digging out stumps or rocks.

Garden rake. Used for smoothing soil, mixing in amendments, spreading mulch. Lawn rake has long, fanned tines good for raking leaves or grass clippings.

Cultivator. Good for raking out shallow weeds, breaking up crusted soil, and working amendments into surface soil.

Lawn aerator. Opens small holes in lawns, reducing compaction and thatch buildup. Core aerators, which remove small cylinders of soil, are better than spike aerators, which push spikes into soil. For large lawns, rent a power model.

Types of Hoes

Some people think of hoes as multipurpose tools, but they fall into two distinct categories depending on their design. The common, large-bladed hoe is an excellent earthmoving tool, but it's more cumbersome for weeding than other styles. Hoes designed for weeding have smaller blades that are easier to slide under the soil and between closely spaced plants. They must be kept quite sharp to work efficiently. A wide variety of hoes have evolved over time to suit different working styles and positions; only the most common types are described here.

Standard garden hoe/American pattern hoe. Used for digging, chopping, moving dirt, creating broad furrows or raised beds, laying out rows.

Grub hoe/Eye hoe. Medium-length handle. Used for digging hard ground, working among dense roots, removing stumps or rocks. Swing it for extra leverage.

Oscillating stirrup hoe/Action hoe. Blade is sharp along both edges, attached so it can move back and forth; most models have detachable blades that are easy to replace. Excellent for weeding in loose soil. Push it forward and then backward to slice off weeds in both directions.

Colinear hoe. Wide, shallow blade is parallel to ground when handle is held at hoeing angle. Excellent for weeding; small blade slides below soil surface to slice off weeds. Buy models with a detachable blade.

Onion hoe/Swan neck hoe. Excellent for weeding; small blade slides below soil surface to slice off weeds. Blade sizes range from 6 to 9 inches wide.

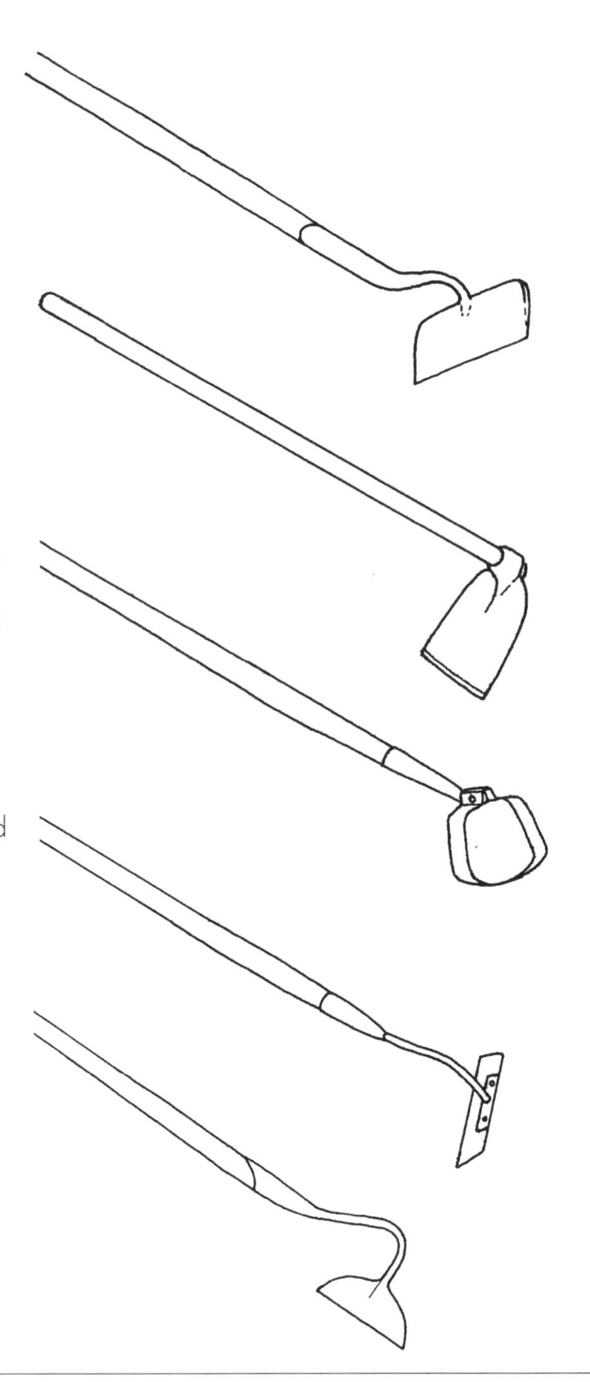

Warren hoe/Gooseneck hoe. Easy to maneuver in small spaces; excellent for weeding or cultivating between plants; also good for digging seed furrows, digging out weed clumps.

Push hoe. Good for slicing off weeds below soil surface and for edging beds. Blade cuts on push stroke. Scuffle hoes cut on both push and pull stroke; the Swoe has a tilted, offset blade that cuts weeds without disturbing mulch.

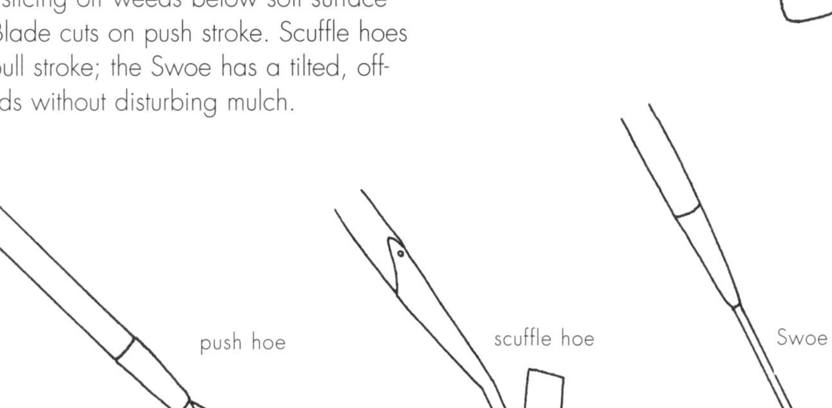

Wheel hoe. Used to weed or cultivate accurately between long rows, or to work in powdered amendments; there are attachments for seeding, hilling, raking, and other tasks.

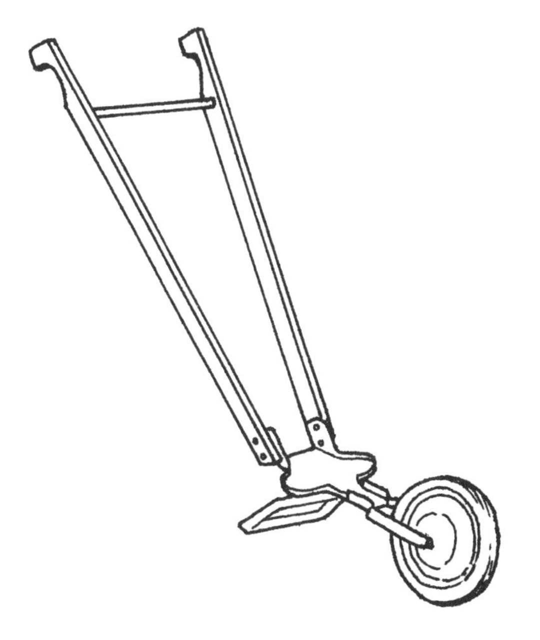

HINTS FOR SUCCESS

Using Hand Tools Safely

- Round up and store all tools at the end of each gardening session. You'll reduce the risk of tripping over them or of running over smaller tools with a lawn mower. You'll also prolong tool life, as rain quickly causes rusting.
- If you must set aside a tool for a moment, lay it with points or curved side facing down. This prevents the rude surprise of stepping on the tool and flipping it into your face. Avoid driving a garden fork into the ground, because it may fall onto your garden plants.
- Replace handles when they break rather than trying to mend them. Replacement handles are available at hardware stores and garden centers. Prevent splinters by sanding any rough areas on wooden handles as soon as they appear.

Sharpening Your Tools

A dull shovel or hoe takes much more effort to use than a sharp one. Digging or dirt-moving tools need touching up only once a season; weeding hoes need resharpening after several hours of use. To minimize sharpening, try to avoid using your weeding hoe for other tasks. Get a sturdy, wide-bladed hoe for moving dirt around or digging furrows, which don't require a super-sharp blade. You can use these same steps for sharpening the blades of a rotary lawn mower; remove the blade from the mower before sharpening.

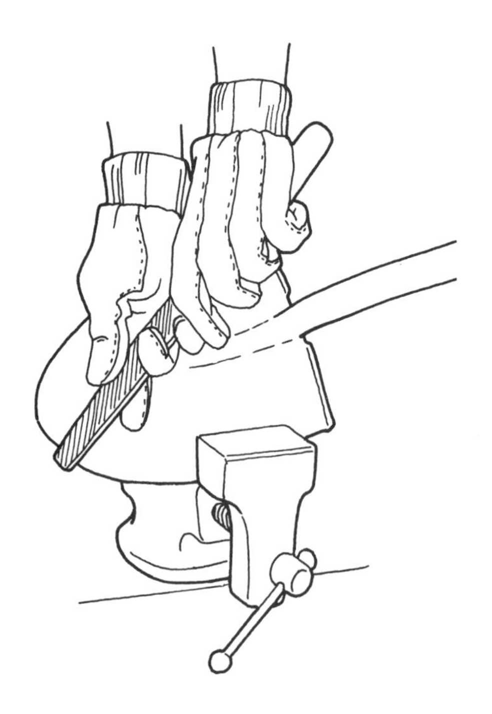

1 Secure the tool in a vise with the dull edge at a convenient height and angle. For shovels and spades, face the front or bowl of the tool toward you. For hoes and weeders, face the angled cutting side toward you (on pull types, this is the side facing you when you pull) and the flat, unsharpened side away from you.

HINT FOR SUCCESS

Which Is the Working Edge?

The working edges of gardening tools are different from those of ordinary knives. One side of the sharpened edge is flat and never needs sharpening. Only the angled (beveled) side is sharpened. Look carefully to make sure you're about to work on the angled side, not the flat side. It's important to try to maintain this existing angle as you sharpen.

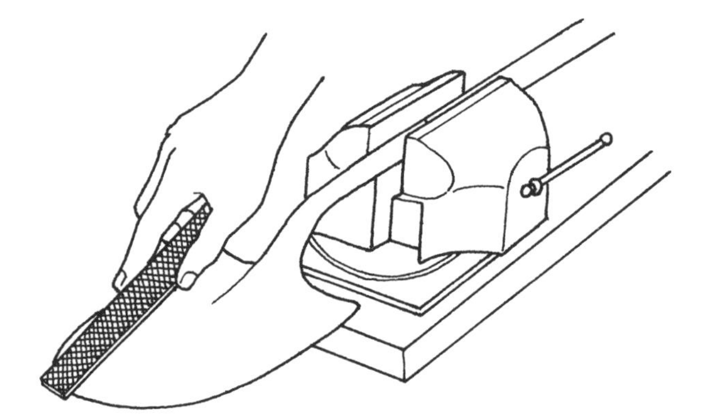

2 Determine the original angle (bevel) of the sharpened edge. Hold the file parallel to the bevel and turn it until you're working at a 30- to 40-degree angle. Pushing away from you with medium pressure, file each dent or nick with short strokes. File in one direction only until the working edge is smooth and even. Remove any sharp burrs on the back side with one or two file strokes; don't sharpen the back side.

Maintaining Hand Tools

A few simple procedures will keep your tools in good shape for many years of use. The same tips apply to long-handled and small hand tools, as both are usually metal blades attached to wooden handles. In addition to the steps below, see the page at left for how to sharpen edges. You'll need much less effort to dig or weed if you keep working edges sharp.

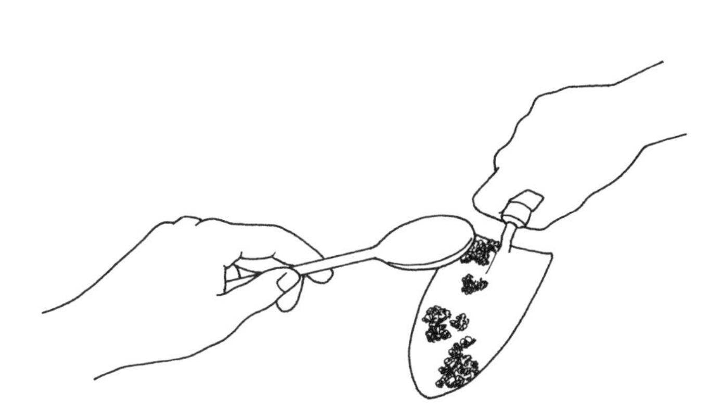

Clean tools after each use. Wipe or brush dirt off, wash it off with a hose, or simply scrape it off with a piece of wood. Some gardeners keep a narrow scrap of wood with one angled end (or an old wooden spoon) just for this purpose. You can also plunge blades into a bucket of clean sand.

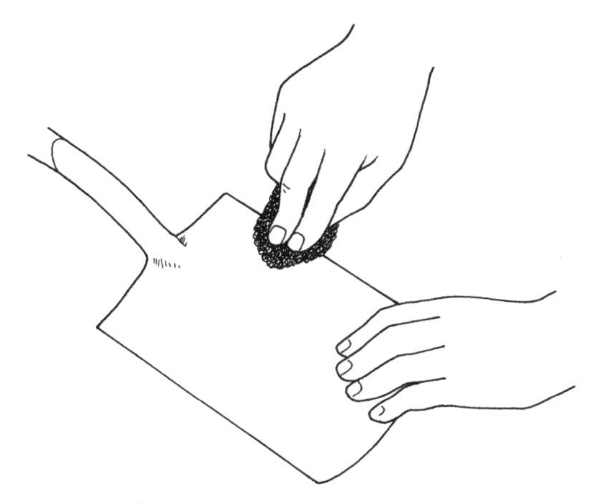

At the end of the gardening season, remove any rust from metal parts with steel wool, sandpaper, or a wire brush. Wipe the cleaned metal parts with an oily rag before storing; a thin coat of oil will help prevent rust.

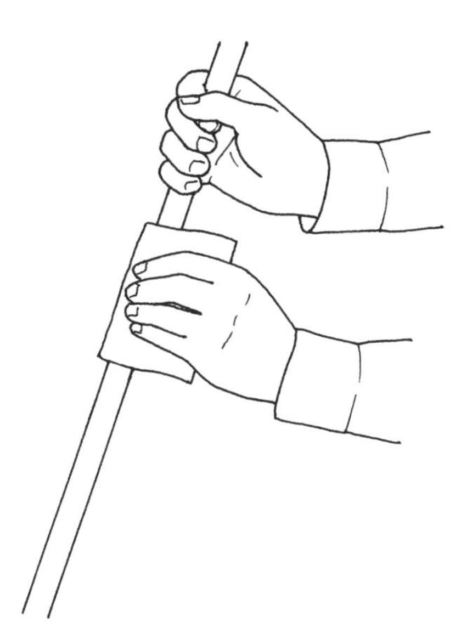

Whenever a wooden handle becomes rough, smooth it with sandpaper to prevent splinters. Wipe the handle with tung oil or vegetable oil after sanding. You may also wish to oil handles lightly before storing for the winter.

Short-handled tools are easy to misplace in the garden. Paint handles a bright color with heavy-duty enamel to make them easy to find. A strip of colored tape around the handle serves the same purpose but needs replacing more often. You can also dip handles in a can of (liquid) plastic handle grip; it dries to make a nice rubbery grip over the wood.

Protecting Your Back

If you sit at a desk all week and then try to shovel a ton or so of manure or compost on Saturday morning, you risk injuring your lower back. Back pain is easy to prevent if you learn to work with your body rather than against it. Take periodic stretch and rest breaks. By giving muscles a few minutes to recover, you greatly reduce the risk of injury. If you overdo it, stop at the first sign of pain. Immediately applying an ice pack (or a bag of loose, frozen peas!) to strains helps minimize discomfort.

Warm up before gardening, as you would before any other athletic endeavor, by doing the following exercises:

Stretch out your legs. Gently arch your back and then slowly bend forward, stretching toward your toes without straining.

HINT FOR SUCCESS

Creative Thinking

Turning compost with a garden fork is one of the most strenuous gardening activities. Some people use a rototiller to do some of the heavy work. First they spread out the pile until it's a foot or two deep. They run the tiller back and forth until everything is well mixed, and toss the mix back into the bin or reshape it into a pile.

Other Ideas:

- Slowly swing arms from side to side while twisting at the waist to loosen your upper back.
- If your lower back tends to get stiff or sore, lie on your back and bring your knees up to your chest several times. Bring knees up once more and hug them to your chest to gently stretch the lower back.

Hold a long-handled tool, such as a hoe, across your shoulders. Bend to each side as far as possible, and hold that position for five seconds. Be careful to keep your body in good alignment and do not bend so far that you lose your balance. Bend to each side five times.

Grasp the hoe handle with both hands and move it to one side of your body, then to the other. Lift as high to each side as you can, hold that position for five seconds, and then relax.

Back-Saving Techniques

Keep your back as vertical as possible as often as possible. When you're upright, you can use your weight and feet rather than arm and back muscles to push a spade into the ground. Bend your knees before you lift, keeping your back upright. Straighten knees to lift each shovelful (or any heavy object) with powerful leg muscles, not weaker back muscles. Keep the load close to your body. Use a cart to move heavy materials any distance.

Never bend while lifting something with a long shovel and then twist to toss the shovelful. This is one of the most stressful actions for the back. Using the more efficient vertical position, the same back can easily lift heavier objects without strain.

Hoes are designed to minimize back strain. Stand erect when hoeing or raking. Hold the hoe's blade parallel to the soil and drag it just below the surface to slice off weeds below their crowns. Most hoes are designed to be pulled toward you. Hold such hoes and rakes with thumbs up, as you'd hold a broom. Try for a sweeping rather than a chopping motion. Chopping with a hoe isn't efficient.

If you're stooping and getting a stiff back, refine your technique or try a different style of hoe.

Vary your tasks to give different muscles a break. Alternate shoveling with a less strenuous activity that keeps your back straight, such as minor pruning at chest level or higher. Lots of pruning can strain your wrists. Cushioned grips help reduce wrist strain; inexpensive ones can be added onto many tools.

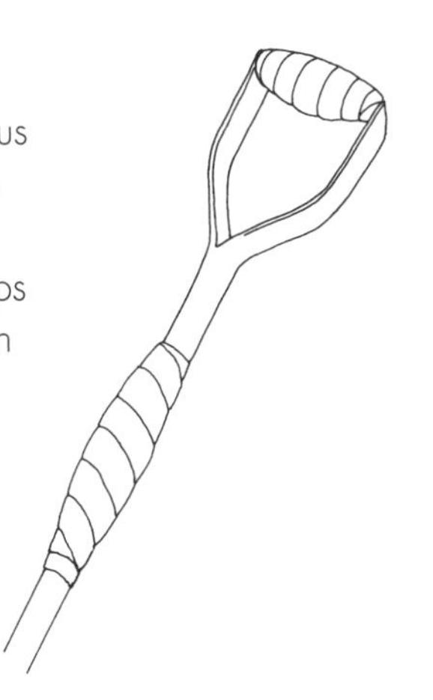

HINT FOR SUCCESS

Back-Saving Tools

Rakes and some other long-handled tools are now available with bent handles. These allow you to stand straighter as you work, reducing back strain.

Power Tools

Some of the gardener's tasks are made much easier with the use of power tools — especially those that process organic matter. Here are some important facts about the most useful types.

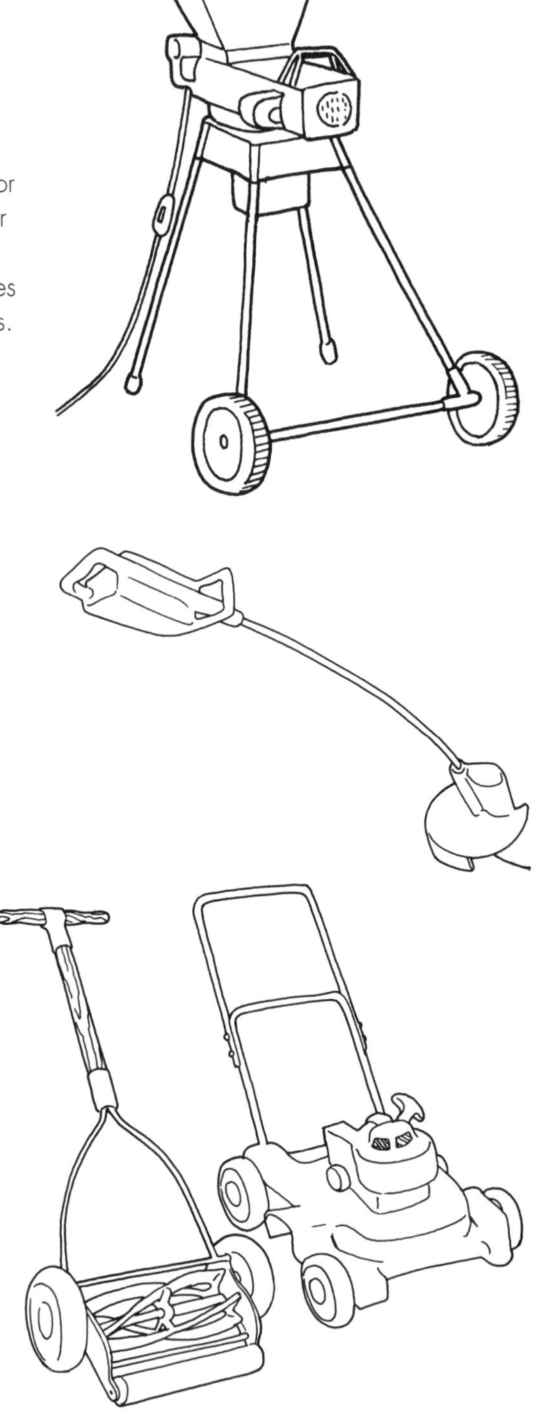

Chippers and shredders. Chippers have one or more knives to slice woody material into small pieces. Shredders pulverize with hammers or other mechanisms; they work best with softer material. Many models carry out both functions. To shred leaves without frustration, buy a machine designed specifically for that purpose.

Large models (8 to 10 horsepower with chipping tubes 4 inches or wider) usually take branches up to 3 inches in diameter, while smaller ones take only 1 ¼-inch branches. Replace or sharpen blades often.

Chip woody material while still green and prune off side branches before chipping. Place machine on a tarp for easy collection of chips.

String trimmers. A trimmer has a small motor on one end that drives a spinning, sturdy nylon line. The cord spins so rapidly that it easily shears off plants. Use it to mow very long grass or weeds and to trim lawns around stones. (Avoid trimming right next to trunks, as you could injure the plants.) It is also excellent for chopping taller green manures before turning under. Gas models cut heavier weeds than most electric ones.

A few models automatically dispense line as it wears off, but most require bouncing the machine's head on the ground to release replacement line.

Lawn mowers. Mowers are no longer simply lawn-cutting tools. Newer models are composting machines that chop grass finely so that it decomposes in place, returning nutrients to lawns. Bagging models are still useful for collecting grass clippings for the compost heap. Any mower can chop leaves for compost or mulch; just run over small, raked piles.

Master Gardening Tips

Using Power Equipment Safely

The same engines that save so much labor can also inflict much damage. All users of power equipment need to be familiar with simple safety precautions. They're just common sense, but they suddenly make a lot more sense when you know people who've hurt themselves, or who work in an emergency room.

- ***Dress appropriately.*** A long-sleeved shirt and long pants help protect your arms and legs. Make sure they're comfortable but reasonably tight fitting, so there's nothing to get caught in moving blades. Since sneakers or running shoes offer little resistance to these machines, wear sturdy shoes or boots with socks. Work gloves are important when using chippers and shredders.

- ***Remember that it can take years*** for subtle hearing loss from power tools to show up, so be sensible and wear ear protectors. As a reward, you'll find mowing the lawn and similar tasks much less stressful. Protect your eyes from flying debris with goggles. Eye protection is absolutely essential with chippers and shredders.

- ***Always turn off the motor*** (unplug electric tools) before attempting to unclog or fix an engine. Before buying a new piece of equipment, make sure it has some sort of safety mechanism that shuts it down when you let go. This can prevent a tiller or lawn mower from running over you if you slip. String trimmers should stop spinning as soon as you release the trigger.

- ***Keep children, spectators, and pets a safe distance away.*** Lawn mowers, chippers, and shredders can throw stones a distance of several yards. Remove branches and similar obstacles from the work area before you turn on equipment.

- ***Allow engines to cool a few minutes*** before refilling with gas; gas spilled onto a hot engine could ignite. Use a funnel whenever pouring gas or oil to minimize spills on your skin, your tools, and the ground. It takes only a few teaspoons (ml) of gas or oil to contaminate water supplies or soil. Dispose of old gasoline and used oil properly; take oil to a garage for recycling. Label all oil and gas cans clearly and keep out of reach of children. Labeling containers of oil/gas mixtures will prevent the expensive mistake of using them instead of straight gasoline.

- ***Plug electric tools only into outlets*** with a ground fault circuit interrupter (GFCI). These specialized outlets shut off the power anytime there's a short circuit, greatly reducing your risk of serious shock. Check power cords before and after each use; replace damaged cords before using.

Before You Buy

These machines can be powered by gasoline or electricity. Electric models are quieter and lighter, and require less maintenance. A few electric trimmers and mowers have rechargeable batteries, eliminating the need for cords.

Rototillers and Cultivators

Rototillers and their smaller cousins save effort in all the tasks that involve digging or mixing soil.

If you're cultivating over half an acre, a large machine is worthwhile, but most gardens don't need the large, expensive models. The heavier the equipment, the greater the risk of compacting soil. Once your garden is established you'll probably find that a smaller, lighter, less expensive model is more fun to use and better suits your needs.

> **Before You Buy**
>
> Consider renting or borrowing a large machine if you're starting a new garden. If you remove sod first and compost it (a good idea in any case), you can use a smaller model even for a new garden.

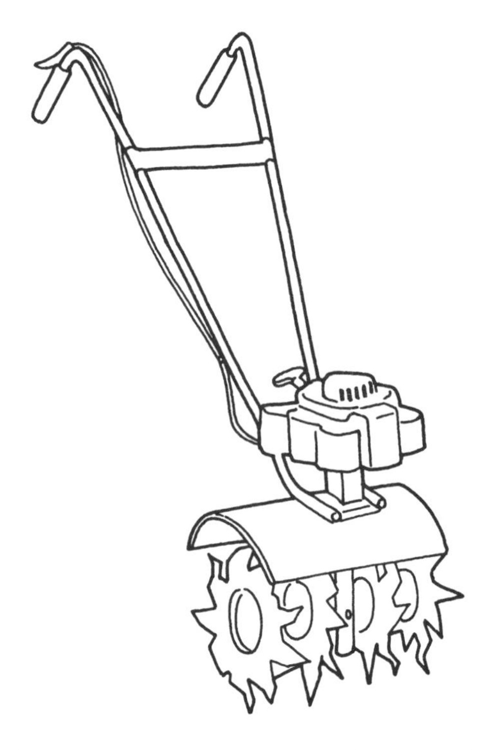

Cultivator. Small, lightweight machine, usually 1 to 2 horsepower (hp); tines are directly below engine on most models. No wheels; user pushes or drives the rotating tines through the soil. Most are gas-powered; a few electric models are available. Tilling widths average 6 to 10 inches. Optional attachments include edgers, lawn aerators, and lawn dethatchers. Least expensive.

Good for weeding and loosening soil to 4 or 6 inches deep and for mixing fine amendments to that depth (effective depth usually less than manufacturers claim). Easy to maneuver around existing plants.

Front- and mid-tine tillers. Larger than cultivators, but relatively small and lightweight, with gasoline engines up to 4 or 5 hp; tines are directly below or slightly in front of engine. User pushes or drives the rotating tines through the soil. Cultivating width is usually 12 to 18 inches. Optional attachments may include one or two detachable wheels, edgers, lawn aerators, lawn dethatchers, and a hilling/furrowing device. Price varies with size.

Good for weeding or loosening soil up to 8 inches deep, and turning under organic matter and green manures. Can handle coarser material than cultivators.

HINT FOR SUCCESS

Attention, Perfectionists:

More is *not* better when it comes to rototilling. You can actually destroy the crumbly structure you've worked so hard to create by overusing these machines. Compaction of subsurface soil from overtilling is so common that it's got a name of its own: plow pan or tillage pan.

Use the fewest passes possible when weeding, digging, or incorporating. Switch to a garden rake for the finishing touches of leveling beds and creating a fine seedbed. To reduce the risk of compaction, never till wet soil — wait until soil is moist (when a handful squeezed in your hand crumbles). (See "When to Work Your Soil" on page 10.)

Rear-tine tillers. Larger, heavier, gas-powered machines; 5 hp is common, but 8 and 12 hp models are available. Wheels are powered and drive the machine in a straight line, making these easier to use than front- or mid-tine tillers if gardens are large. Cultivating width averages 18 inches, but adding tines can increase this to 25 or 30 inches. Expensive.

Good for weeding only in very large gardens where rows are widely spaced (hard to turn in small beds). Excellent for turning under amendments, organic matter, and green manures; larger engines can break sod and hard or rocky ground.

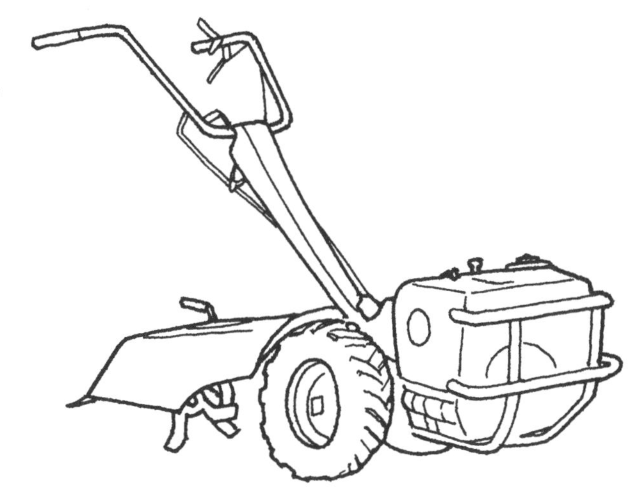

Tractor-mounted tillers. Heavy machines that are either powered by the tractor or self-powered; tilling widths are 20 to 48 inches (even wider on a few models). Expensive.

Perform all of the same functions that other tillers do. Make sense only for large gardens and people who already have lawn or garden tractors. Extra weight increases the risk of compacting soil.

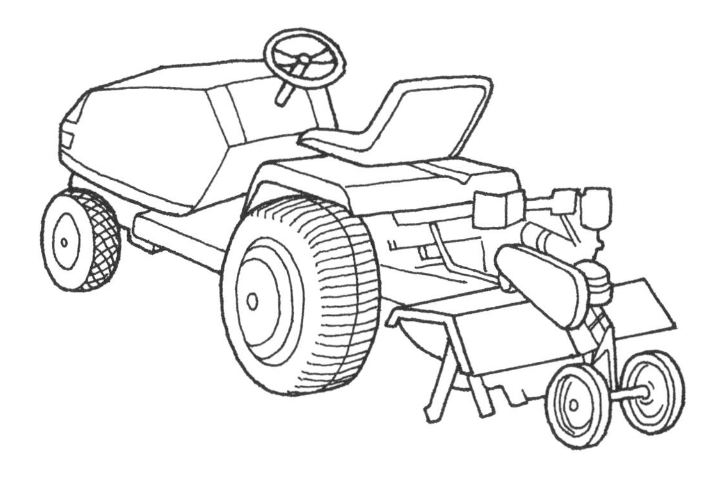

Maintaining Power Tools

Basic maintenance is even more important for power tools than it is for hand tools to assure a long, useful life. A few simple steps should keep machines running for at least 15 years without an overhaul. (After 15 years, a professional overhaul can double a tool's usable life.) Servicing small engines requires no special tools or skills.

With average use, the following fall tune-up should be all that's needed, aside from checking the oil (and gas!) level before each use. Heavy use may require more frequent oil changes; follow the manufacturer's recommendations. Occasional cleaning is also a good idea. Brush, wipe off, or clear machines whenever the outsides or moving parts become caked with plant debris or mud. Clean soil off tillers after every use.

HINT FOR SUCCESS

If your spark plug is in good shape, your engine is running well. Look at the plug's threaded end. The ceramic insulator should be light brown to gray. If the insulator has burned to white, or if the plug end is covered with soot or black oil, take your engine in for a professional tune-up.

MATERIALS

- Adjustable wrench
- Shallow pan (capacity at least 1 quart or liter; dispose of when finished)
- Gasoline stabilizer (optional)
- Funnel (disposable)
- Jar (for used oil)
- Fresh oil (check manual for weight)
- Spark-plug wrench
- Old toothbrush
- Old tablespoon (use only for oil)
- Small plastic bag and bucket of hot, soapy water for foam air filters
- Replacement filter for paper air filters
- Paper towels
- Oil can or spray lubricant

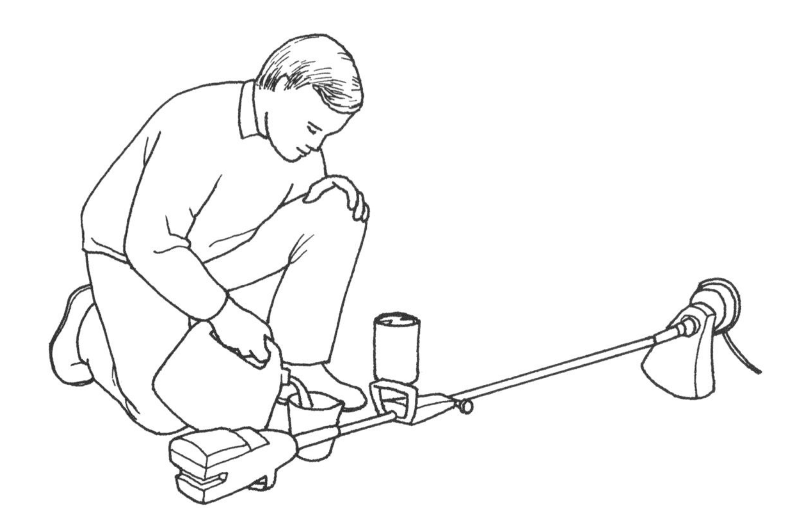

1 Run the engine to use up all the gas before storing tools for winter. Or add gasoline stabilizer (available at auto parts stores) and run the motor briefly to mix it in. Either is easier and safer than draining the gas; both prevent engine clogging. Buy fresh gas next year.

String trimmers and other tools that use gas mixed with oil should be left empty whenever they won't be used for a month. Mix only small amounts of oil and gas at one time; discard the mixture after eight weeks. Store in a tight, labeled container.

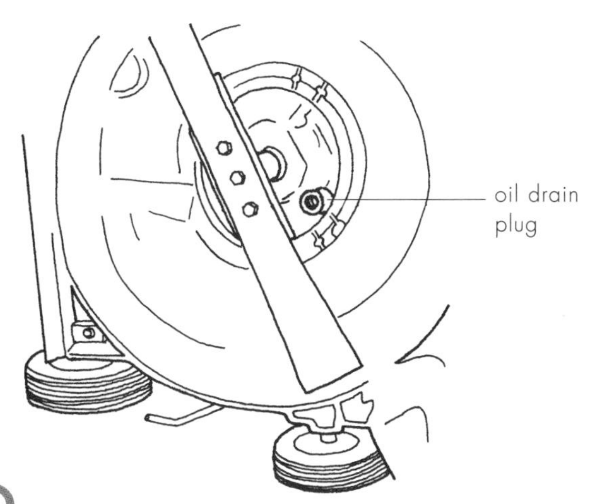

2 While the engine is still warm from use, drain and replace the oil. First remove the wire from the spark plug for safety. Tilting the engine if necessary (keeping the muffler side up), loosen the oil drain plug slightly with a wrench. Place the pan below the oil plug, unscrew the plug, and let the dirty oil drain into the pan; replace the plug. Refill the engine oil as recommended; avoid overfilling. Reconnect the spark-plug wire before storing or using.

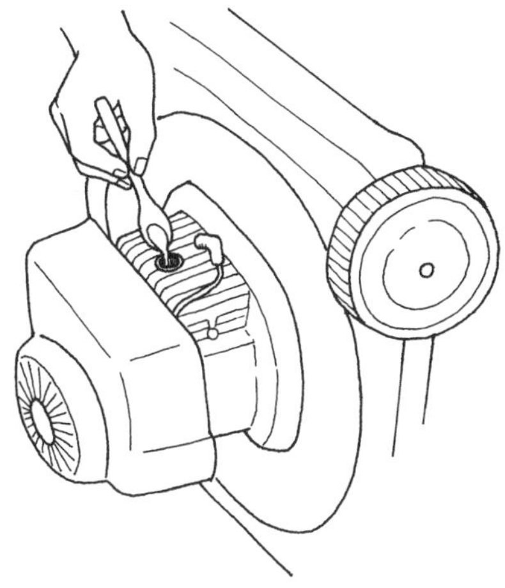

3 A bit of oil inside protects the engine from winter corrosion. Remove the spark plug with a spark-plug wrench and clean with an old toothbrush, or buy a new spark plug (usually after every 50 hours of use). Pour a tablespoon (15 ml) of clean engine oil into the hole. Replace the spark plug and crank the engine over by hand a couple of times to distribute the oil evenly. When you start the engine next spring you'll see a bit of white smoke as the oil burns off.

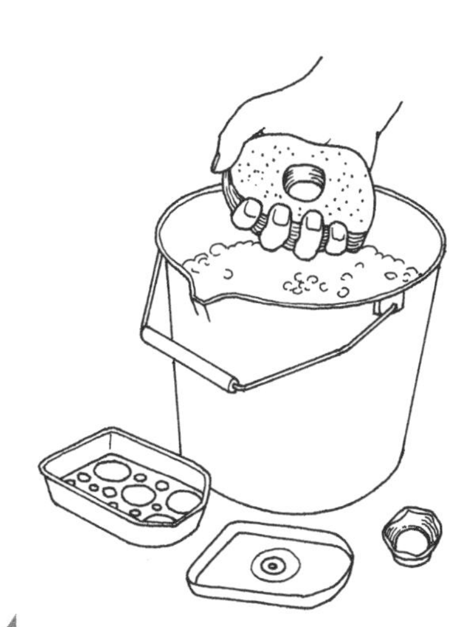

4 Clean the air filter to ensure a good air supply to the carburetor. Remove the cover and remove the filter underneath. Replace paper filters each fall for best performance. Wash foam filters in hot, soapy water. Squeeze as much water out as possible. Place 3 tablespoons (45 ml) of your engine oil in a small, clean plastic bag. Put the cleaned filter in the bag, close tightly, and squeeze the filter repeatedly until it's saturated with oil. Remove any excess oil with a paper towel before replacing the filter.

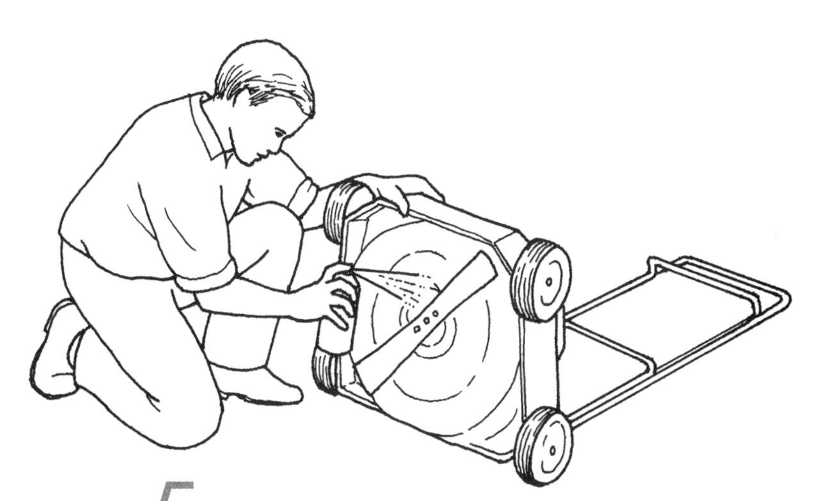

5 Before storing, brush any debris off the engine, especially around the cooling fins. Inspect the starting cord and replace if it looks frayed. Find the fuel line and bend it gently to see if it has developed any cracks; if so, replace. Sharpen blades or tines. Use a bit of lubricating oil or spray on all moving parts, including the control cable and carburetor linkage. (Don't go overboard; too much lubricant attracts dirt.)

Storing Tools

Smart storage makes it easy to find tools the instant you need them. You won't have to sort through a jumble or worry about heavy blades falling on you as you root around. Keep tools out of reach of the elements, and keep them up off an earth floor to reduce rust and extend their useful life.

Inventory your tools before designing a storage area. Lay long-handled tools side by side to measure how long a board or rack you'll need for hanging them. Do the same for your short-handled tools. If you're building a shed, make sure it's not only big enough to hold everything but also allows you to walk around carts and large power tools. You need to be able to get *to* all your tools.

Where space is limited, prioritize. Save the best spots — those within easiest reach of the garden — for the tools you use most often. Rarely used tools or amendments, or those that won't be used in the current season, can be relegated to less accessible spots.

A Store fertilizers in moisture-proof containers such as large plastic tubs with tight-fitting lids. Many synthetic fertilizers are notorious for absorbing water from the air and changing from dry, powdery crystals to a sticky mess. Tight containers are a good idea for organic fertilizers, too, as they keep animals from smelling possibly tantalizing ingredients. (A sturdy cupboard with a snug door will also keep animals out.)

B Keep bags of amendments off the ground. If they're too large to fit in tubs, set them on short pieces of 4 x 4 lumber or upside-down plastic milk crates. The space below a hanging cupboard or potting bench is often perfect for bags of amendments.

C Hang long-handled tools on a wall so that you can grab one without having to sort through a whole tangle. Mount a length of 2 x 4 board at a convenient height and drive pairs of nails to support blades or D-handles, or install sets of individual handle-gripping clips. You can buy a variety of ready-made tool racks. Just make sure any system allows you to remove one tool without disturbing the others.

D Store short-handled tools in one accessible spot. Options include hanging on a rack (perhaps above a potting bench), laying out on a shelf, and storing in a hanging cupboard. If you hang them from a pegboard, draw an outline around each tool so you know if one's been left out in the garden.

E You can buy or make a many-pocketed fabric holder that fits around a 5-gallon tub. This not only organizes storage for small tools, it also gives you a handy carry-all for carting them around the garden.

F Exposed 2 x 4s are great for mounting shelves, racks, and specialized tool holders. To make your own hose holder, secure two 1-foot (30 cm) lengths of scrap 1 x 3 board to either side of a vertical 2 x 4 using four screws. A convenient height is about 3 feet (90 cm), just above your waist.

G Keep the most frequently used power tools accessible. Store the lawn mower in front for the summer; if your shredder is used only for fall or spring cleanup, it doesn't need to hog a prime spot the rest of the year. A string trimmer can hide behind a snowblower (or in an out-of-the-way corner of a basement) all winter.

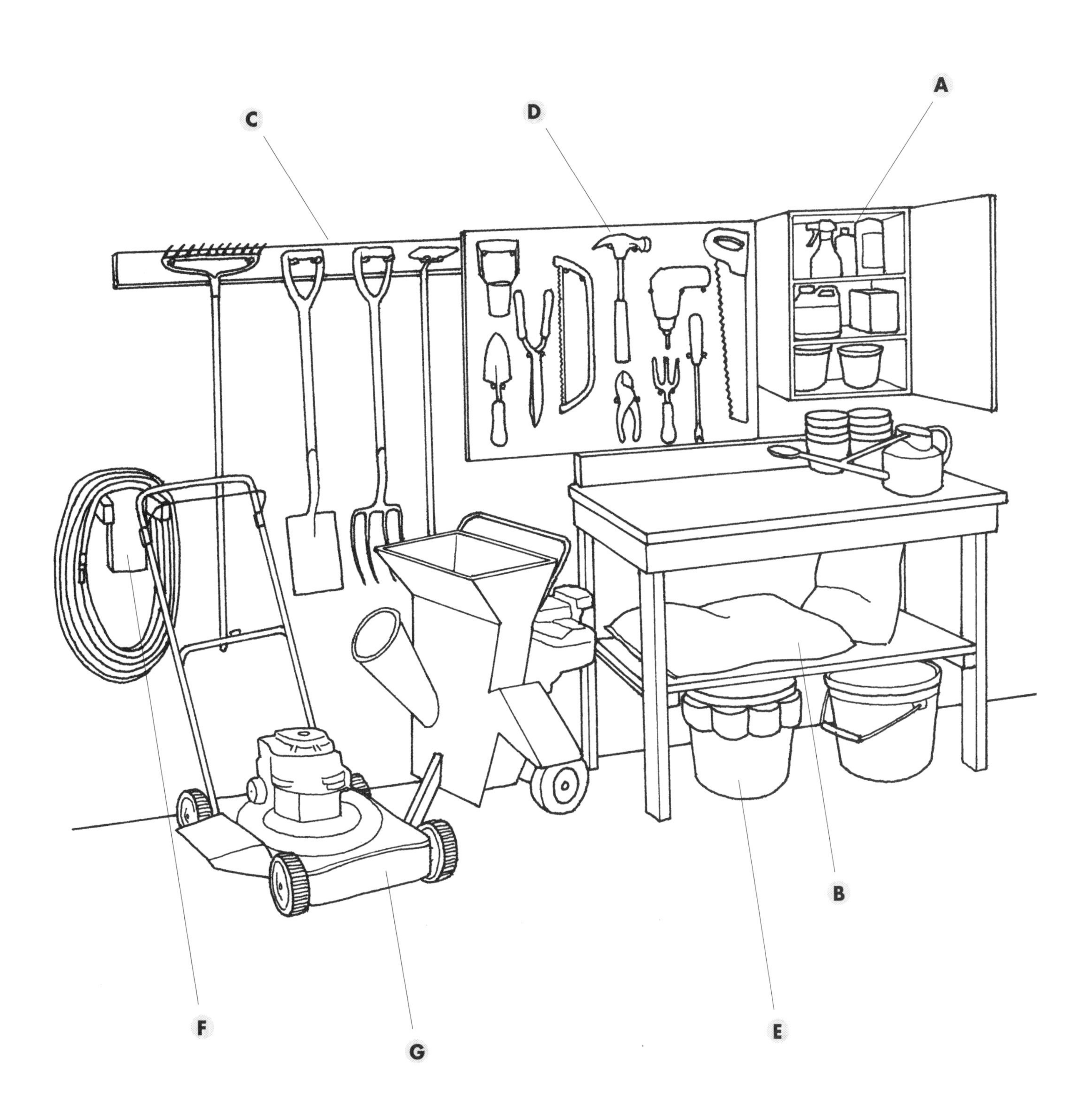
A
B
C
D
E
F
G

Handy Haulers

Every gardener needs something for hauling — for compost, for amendments, and for major weeding or clean-up sessions. While a variety of baskets and buckets come in handy for small loads, large and heavy loads require wheels.

Wheelbarrows. Best for moving small loads of sand, gravel, bark chips, and dirt. Single wheel in front makes them easy to turn in small spaces, but likely to tip sideways.

Garden carts. Excellent for hauling heavy and bulky loads; very stable; larger models are difficult to turn in tight spots.

Front panel removes for dumping. A heavy-duty aluminum model is now available that folds flat for storage. Heavy-duty plastic hybrids are easy to hose clean.

Tarps. For moving large piles of leaves, grass clippings, and similar light but bulky items; also to cover compost piles. Specialized, small tarps have handles at each corner or corners sewn into "bags" with handles.

Baskets and buckets. Handy 5-gallon tubs can be recycled from contractors who use a lot of Sheetrock compound. Tubs with tight-fitting lids are good for storing fertilizers or amendments. Garden baskets should have sturdy, reinforced handles.

CHAPTER 4

Making and Using Compost

In This Chapter

- Not-So-Secret Ingredients
- Siting Your Bin or Pile
- No-Fuss Cold Compost
- Is It Done Yet?
- Making the Ideal Hot Pile
- Tips for Faster Compost
- Survey of Composting Bins
- A Garbage-Can Composter
- A Simple Large Bin
- Triple Bin: Cadillac of Composters
- Composting with Earthworms
- Outdoor Worm Bins
- Sheet Composting
- Pit or Trench Composting
- Using Your Compost
- Compost Troubleshooter

Compost is the best form of organic matter to add to your soil. It's a mix of organic materials, such as grass clippings, garden waste, and kitchen scraps, which have decayed into a crumbly, dark mass. A balanced, slow-release source of nutrients, compost can keep soils stocked with all the micronutrients plants need. It supplies small amounts of major nutrients and helps soil hold onto nutrients and water long enough for plants to use them. It increases the overall health of your garden by suppressing disease organisms in the soil. When sprayed on leaves in the form of compost tea, it can suppress leaf spots, mildews, molds, and other leaf diseases. Compost also encourages beneficial soil organisms, which feed on disease organisms or make nutrients available to plants.

Compost is so simple that you might as well make it yourself instead of buying it. After all, you're just trying to speed up nature's nutrient recycling program, which goes on constantly all around us. There's no need for an elaborate system. Simply piling leaves will do if you're not in a rush. If you don't have deciduous trees, you can make concentrated compost — pure humus — from kitchen scraps with a low-maintenance indoor worm bin.

If you want to make compost quickly, build a "hot," or active, pile. Under ideal conditions, the organisms responsible for decomposition will experience a population explosion, generating a lot of heat. You need to mix the right balance of materials that are carbon-rich and nitrogen-rich and turn piles often to give the organisms enough oxygen. If you'd rather not bother with hot piles, you can make great compost using "cold," or passive, piles. Just pile up whatever materials you have on hand and turn the piles rarely or not at all. (To speed things up, turn often and chop all ingredients before piling.)

Not-So-Secret Ingredients

Many materials are good for composting. Including a variety helps ensure a well-balanced supply of micronutrients. Remember that tough materials such as sticks and cornstalks take much longer to break down than softer materials, unless you chop them into small pieces.

Balance Browns and Greens

The best compost is made from a mix of "browns" and "greens." Browns are dry or dead plant materials such as fallen leaves, straw, and wood shavings; they supply lots of carbon. Greens are fresh plant materials and fruit and vegetable scraps from the kitchen, regardless of color, that supply nitrogen and protein. The best sources of protein are animal by-products, such as manure, milk, blood, and wool; these don't fit into a color category.

Microorganisms

The microorganisms that produce compost need a balanced diet of carbon (their energy source) and nitrogen-rich protein. Unless you use materials that provide a balance of these (see chart on page 57), top each 6-inch layer of carbon-rich material with a 2- to 4-inch layer rich in nitrogen. Or, add about a pound of nitrogen-/protein-rich material for every 30 pounds of carbon-rich material (about three-fourths of a hay bale). Extra-high-carbon newspapers and fresh sawdust need twice as much nitrogen-rich material to balance their effects. Anything that's already started decomposing has a more balanced content than fresh material and therefore needs less nitrogen. If you have too much nitrogen, the pile could start to smell.

If you don't want to bother keeping track, don't worry. Without enough nitrogen (assuming that no other factors in the "Compost Troubleshooter" on page 82 apply), your pile will still break down, only more slowly. It also won't heat up.

Boost Nutrient Content

Some ingredients are added to supply nutrients to the soil when you use your compost. Thin sprinklings of wood ashes boost the potassium and phosphorus content. Sprinklings of rock powders such as granite meal and greensand supply potassium and many micronutrients, though very slowly. Don't add limestone (or thick layers of wood ashes, which contain some lime) until your compost is finished. If lime comes in contact with high-nitrogen materials, it creates ammonia gas, and you'll lose nitrogen by evaporation.

MASTER GARDENING TIP

Use Activators

Any substance that speeds up decomposition in your compost pile is an activator. Generally, activators supply nitrogen (protein) or microorganisms, or both. Save money by making your own from the ingredients here. Avoid synthetic fertilizers, which contain no protein and seem to inhibit compost organisms. Activators work best if mixed in thoroughly. If you'll be turning your pile soon, simply spread a thin layer every 6 inches or so. Use only 1 cup of dry meal per layer but up to a 2-inch layer of soil or compost.

Nitrogen/Protein Sources

- Alfalfa meal
- Blood meal
- Dehydrated manure
- Fresh grass clippings
- Fresh manure
- Hoof or horn meal

Microorganism Sources

- Compost (the fresher, the better)
- Fresh or well-aged manure
- Healthy, humus-rich soil
- Strips of sod

What to Use

Nitrogen-/ Protein-Rich ("Greens")	Carbon-Rich ("Browns")	Balanced
Fresh grass clippings	Straw and hay	Manure with bedding
Fresh manure	Leaves	Well-aged manure
Kelp meal, seaweed	Pine needles (slow to break down)	Pea and bean pods
Legume plants (peas, beans, etc.)	Cornstalks	Fruit peels, cores
Alfalfa hay or meal	Sawdust	Vegetable peelings
Crushed eggshells	Wood shavings	Sod and soil
Cabbage leaves, broccoli	Shredded newspaper	Fresh weeds
Sour milk	Dry grass clippings	Rotted wood
Apple or winery pomace	Dry, brown weeds and garden trimmings at season's end	Soft, green garden trimmings
Blood meal	Rice or cocoa hulls	
Cottonseed meal		
Wool		
Soybean meal		
Human hair		
Coffee grounds		

What to Avoid and Why

Material	Reason
Diseased plants	May spread the disease if compost doesn't get hot enough.
Weeds with seeds, or weeds that can sprout from bits of root	Seeds or bits of root may survive and sprout in garden if compost doesn't get hot enough to kill them.
Fresh sewage ("biosolids"), pet feces, used kitty litter	May carry parasites and diseases that infect humans.
Nonbiodegradable items (glass, synthetics, pressure-treated wood)	Will not break down, so you'll just have to pull them out of finished compost.
Toxic chemicals (pesticides, etc.)	Could kill composting organisms.
Charcoal (as in briquettes)	Does not break down in compost.
Coal, coal dust, coal ashes	May contain levels of sulfur or iron that are toxic to plants.
Fats, oils, grease	Large amounts attract animals and keep anything they coat from breaking down.
Meat scraps, bones, cheese	Large amounts attract animals and are very slow to break down.

Composting Systems at a Glance

Hot composting and cold composting are the main systems; all the rest are variations. Bins or boxes and tumblers are simply containers for making hot or cold compost; many appear under "Survey of Composting Bins" on pages 66–67. Hot compost requires a minimum volume: Bins or piles must be at least 3 feet by 3 feet and filled to about 4 feet to allow for settling. Bins and tumblers that hold less than a cubic yard are only good for cold composting. The last three systems are easy alternatives to piles; all three are variations on cold composting.

Type	Advantages	Disadvantages
Hot (fast) outdoor pile	Fast results; kills weed seeds and pathogens; more nutrient-rich than slow (cold) piles because there's less leaching of nutrients; less likely to attract animals or flies. Also called active composting.	Requires careful management and lots of effort to turn and aerate; works best when you add a lot of material at once.
Cold (slow) outdoor pile	Easy to start and add to; you can add material continuously, a little at a time; low maintenance; resulting compost is especially rich in beneficial soil organisms. Also called passive composting.	Takes a year or more to decompose; some nutrients are lost to leaching unless pile is covered; can attract animals and flies; doesn't kill weeds or diseases.
Bin or box	Neat appearance; holds heat more easily than free-standing piles; deters animals if covered; lid keeps rain from leaching nutrients. Can use for hot or cold piles.	Requires time to build or money to buy; decomposition is slow unless you turn materials inside bins; may need to add water from time to time.
Tumbler	Neat appearance; can produce compost quickly if ingredients are chopped finely and balanced; easy to aerate by turning; odor not usually a problem; no nutrient leaching into ground. Good for cold composting (usually not large enough for hot composting).	Relatively expensive; volume is relatively small, rarely large enough to heat up as a hot pile; works best if material is chopped and added all at once; may need to add water from time to time.
Garbage can (or plastic bag)	Easy to do year-round; can be done in small space; requires little effort, though you must poke aeration and drainage holes in garbage can; inexpensive. Only for cold piles.	Mostly anaerobic, so smell can be a problem; can attract fruit flies; you must ensure correct carbon/nitrogen ratio to avoid a slimy mess; doesn't kill weeds or diseases.
Triple bin	Relatively easy to turn material into adjacent bin; one bin can be used to stockpile ingredients while another holds unfinished compost. Accommodates either hot or cold piles.	Best for large volume of material; turning and aerating requires some effort.
Worm composting	Easy; little or no odor; can be done indoors or out in a small space; can be added to continuously; castings are so nutrient-rich they can be used as fertilizer; good way to compost food waste.	Requires some care when adding materials and removing castings; you must protect worms from temperature extremes; can attract fruit flies; may not kill weeds or diseases.
Sheet composting	Can handle large amounts of organic matter; no container or turning required; easy way to improve soil over large areas; boosts earthworm populations.	Materials take several months to decompose; requires effort to spread; doesn't kill weeds or diseases.
Trench or pit	Quick and easy; no maintenance; no investment for materials; boosts earthworm populations; doesn't attract flies or animals; kills many weeds and diseases.	Requires some effort to dig pit or trench; incorporates relatively small amounts of organic matter; improves only small area.

Siting Your Bin or Pile

To start composting in your yard, first choose a good site. Pick a relatively level area so you won't have to haul finished compost uphill to the garden. If you plan to use manure or other heavy ingredients, pick a spot where you can easily dump these nearby. Shade is ideal in hot climates; sun is better in cool climates. Finally, nothing dampens composting enthusiasm like a sour smell wafting through the window. Don't site your bins or piles just upwind from or right next to the house, at least not until you've made a few odor-free batches.

If your site is in a low spot or is poorly drained, elevate the pile on a mound of soil, rot-resistant boards, or a discarded pallet.

DID YOU KNOW?

The Power of Compost: A Case Study

Researchers at the Connecticut Agricultural Experiment Station have been running experiments since 1982 to see how compost affects yields of common vegetables. While it took four or five years for the benefits to show, the results are impressive.

In some soils, compost can completely replace added lime (test your soil periodically to make sure). It also greatly reduces the need for fertilizer. Yields of vegetables from unfertilized soil given an inch (2.5 cm) of compost each year equal those grown in fertilized, but uncomposted, soils. Small amounts of fertilizer (preferably based on a soil test) really boost yields when teamed up with a little compost.

For the experiments, amendments were applied every year. Compost made from oak and maple leaves was spread 1 inch thick on all but one of the beds. The control bed got no compost but received the full recommended rate of fertilizer and lime. Other beds got different amounts of fertilizer and lime, plus the compost.

After 12 years, the yield from the bed that got everything (full rate of amendments plus compost) was 25 percent higher than the bed that got no compost but just the full rate of fertilizer! Another bed got two-thirds the recommended fertilizer, plus compost; it produced similarly higher yields. And the plot that got no fertilizer or lime, just the leaf compost, produced the same amount of vegetables as the control bed!

The soil getting compost had almost twice as much organic matter as the bed with no compost. As a result, the compost-amended soil held nearly a two-week supply of water. This reduces plants' water stress and the need to irrigate. All that organic matter also buffered the soil's acidity. The pH of the bed that got only compost and no lime increased from 5.6 to 6.6, right into the optimum range. Later tests that substituted other types of compost for the leaf compost produced similar results.

No-Fuss Cold Compost

Any pile that doesn't heat up is by default a cold pile. If you don't want to fuss about ingredients and layers or exert the effort to turn piles, you don't have to. Cold compost is great for the garden. It contains an even wider array of beneficial soil microbes than hot compost. It just takes longer to make.

Cold composting works well with any size or type of bin or pile. The simplest form of cold compost is leaf mold. Pile leaves (chopped, if possible) in a corner of the yard and forget about them. In a couple of years you'll have wonderful leaf mold, which is just nature's compost. A fancy tumbler can produce compost relatively quickly, but many are too small for the contents to heat up, so these also make cold compost. Even a well-layered, frequently turned bin may produce cold compost if it has too much carbon, the ingredients are too dry, or rain soaks it.

Bagging It

You can make cold compost anywhere, even in a plastic garbage bag. Unlike the other methods discussed in this chapter, this is anaerobic — oxygen-deprived — composting. The most noticeable difference is that anaerobic compost smells; people who make anaerobic compost unintentionally by creating totally soggy piles find this out. You won't be able to smell anything as long as the bag is sealed. Be prepared for a strong odor when you open the bag, though. Fortunately, the smell disappears quickly once contents are spread out and exposed to oxygen.

Fill a large bag with a mix of chopped leaves, grass clippings, and kitchen scraps. Or add a sprinkle of activator for every couple of shovelfuls of bulky, carbon-rich material until the bag is nearly full. Sprinkle a couple of quarts of water over the contents and mix until all ingredients are moistened. (Shake small or light bags; roll large or heavy ones.)

Tie the bag closed and leave in an out-of-the-way spot where it'll stay above 45°F (8°C) for a few months. For faster results, turn it over every few days by simply rolling it on the ground.

Is It Done Yet?

Compost is finished when it develops the sweet, woodsy smell of rich soil and most ingredients have become a crumbly or fluffy, dark brown, soil-like material. Some large or recognizable pieces usually remain, but everything's a relatively uniform dark brown color. The center of a hot pile is no longer warm, and if you turn your pile it no longer heats up. After you've made a couple of batches, you'll recognize the difference between finished and unfinished compost. If you're not sure, try the following.

A Simple Test

Soak a couple of large spoonfuls of your compost in a cup of water. Fill another cup with plain water (distilled is best, or let tap water sit overnight so chlorine can evaporate). Put 8 or 10 radish or lettuce seeds in each cup and soak them overnight. Dampen two paper towels. Strain off the water and spread the seeds on separate paper towels. Place towels in separate plastic bags and keep warm for a few days until the seeds sprout. (Check periodically to give them some air and make sure they're moist but not soggy.)

Both batches should germinate equally well. If there's a difference of only one or two seeds, it may be the seeds rather than the compost. If significantly more of the seeds soaked in plain water sprout, it shows that your compost needs to sit longer before being used near seedlings. Use unfinished compost only as mulch for shrubs or trees.

MASTER GARDENING TIPS

Why Compost?

- To save on fertilizer costs
- To improve soil tilth and fertility
- To increase numbers and activity of beneficial soil organisms
- To improve soil pH
- To reduce pests and diseases
- To recycle kitchen and yard wastes

HINTS FOR SUCCESS

Collecting Kitchen Scraps for Compost

To make it easy to save kitchen scraps, find a suitable container and keep it in a handy spot.

- Keep a lid on to contain odors and keep out fruit flies and houseflies.
- A handle makes hauling easy.
- Try keeping a bucket of sawdust next to the container and cover each addition of scraps with dry sawdust. This soaks up excess moisture and cuts down on flies and odors.
- If you can stand to make frequent trips to the compost pile, smaller containers are nicer because they get emptied and rinsed more often and therefore are less apt to smell. Large peanut-butter tubs with handles work well. Some people use ceramic containers or cookie jars.

Making the Ideal Hot Pile

Hot composting takes more effort than cold composting, but you'll be rewarded with faster results. Your pile will also generate enough heat to kill pesky diseases and weeds.

To make hot, fast compost, follow all the recommendations under "Tips for Faster Compost" (pages 64–65). In addition, make your pile or bin at least 3 feet square and 3 feet high (4 feet is even better). The site should be well drained. Stockpile ingredients next to the bin or pile so you can assemble as much as possible all at once. *(For metric equivalents, see "Useful Conversions" on page 208.)*

MATERIALS

- Bin or site for pile 3–4 feet x 3–4 feet x 3–4 feet
- Chopped compost ingredients
- Compost activator (see page 56)
- Hose with spray nozzle
- Lid or tarp to cover

1 Spread 6 inches of loose, bulky, carbon-rich material in the bottom of the bin or pile. Cover with a 2-inch layer of fresh nitrogen-rich material; use only 1 inch if the material is dry. Add up to 2 inches of kitchen scraps, weeds, soil, or other balanced material.

2 Unless you used fresh manure for the nitrogen layer, add a handful of fresh compost or good soil (or a strip of sod). For your first batch, you can use a purchased activator to supply microorganisms; a handful of compost from this batch will then supply later batches.

3 Water the layers until the materials are evenly moist but not soggy.

4 Repeat the steps until your stack is 3½ to 4 feet high. Don't worry if it's taller than your bin; the height should drop fairly quickly. Cover with a lid or tarp.

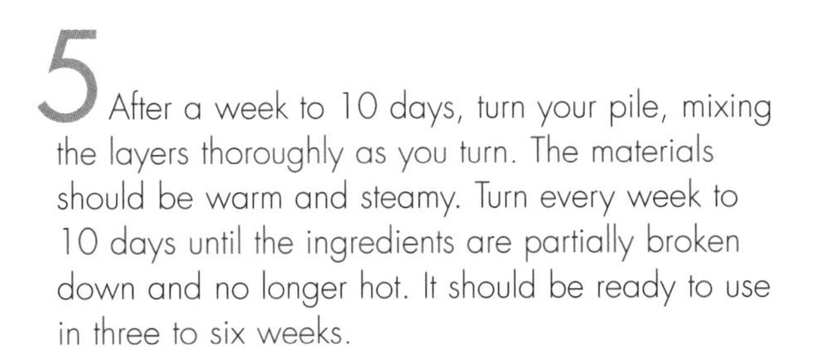

5 After a week to 10 days, turn your pile, mixing the layers thoroughly as you turn. The materials should be warm and steamy. Turn every week to 10 days until the ingredients are partially broken down and no longer hot. It should be ready to use in three to six weeks.

Tips for Faster Compost

Whether you make hot or cold compost, the following tips apply. The more of them you follow, the faster you'll get finished compost. If you're in no rush, you can ignore them. Sit back, relax, and let nature take its time; you'll still get good results.

Start your pile when it's warm outside. Organisms in compost work very slowly when it's cold, and they stop working if they freeze. Warm-season piles are ready sooner than cold-season piles.

Maintain a balance of about 3 parts carbon-rich material ("browns") to 1 part nitrogen-/protein-rich material ("greens"). Cover every 6-inch layer of bulky, loose, brown material with a 2-inch layer of green (you need only 1 inch of a dry, powdered nitrogen source such as blood meal). Or mix 3 parts browns to 1 part greens as you run materials through a shredder.

Turn your piles using a garden fork or aerating tool. Each turning cuts the rotting time by about half. Turning speeds decomposition by mixing and aerating.

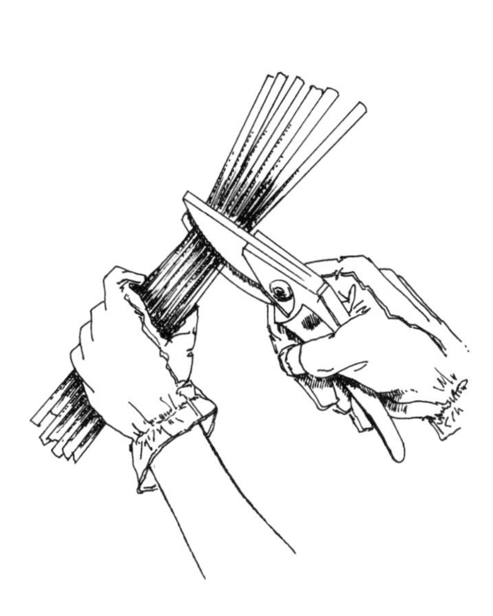

Chop all materials. A shredder is handy for making lots of compost quickly. Otherwise, mow over small piles of leaves to chop them. Cut stiff stems into small pieces with hand pruners; leave out woody material unless you're willing to chop it finely. Break up baseball-bat zucchini with a shovel. Chop kitchen wastes, especially tough rinds (citrus, melon, banana peels), before tossing into your compost bucket. Dry eggshells overnight and then crumble into the compost bucket.

Cover your compost. A tarp or layer of inverted sod helps keep piles and bins moist in dry climates, and keeps rain from soaking them in wet climates. It also prevents nutrients from washing away. If you don't cover your piles, make a shallow depression on top to collect rain in dry climates; in wet climates, round the top to shed rain.

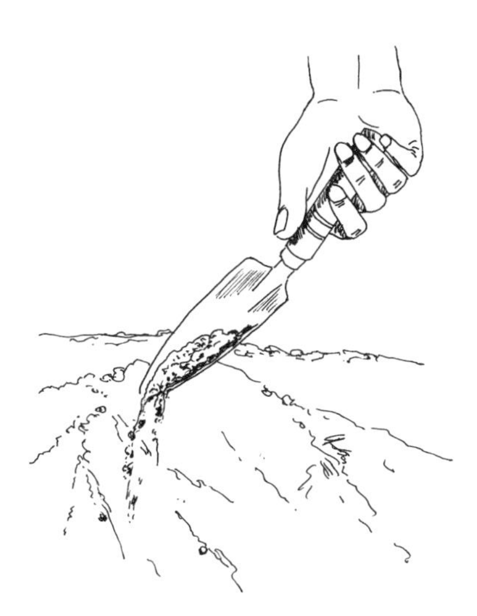

Check moisture levels by digging a 6- to 8-inch hole in the pile and feeling the edges with your bare hands. Piles that are soggy or dry take much longer to break down than those that are evenly moist. Materials in the pile should feel moist to your fingers but not so wet that you can squeeze out moisture.

Aerate your piles. Try a compost aerating tool (available from gardening suppliers) if you don't have the energy to turn piles. Or, for automatic aeration, lay perforated drainage pipes in layers as you build up your pile.

Survey of Composting Bins

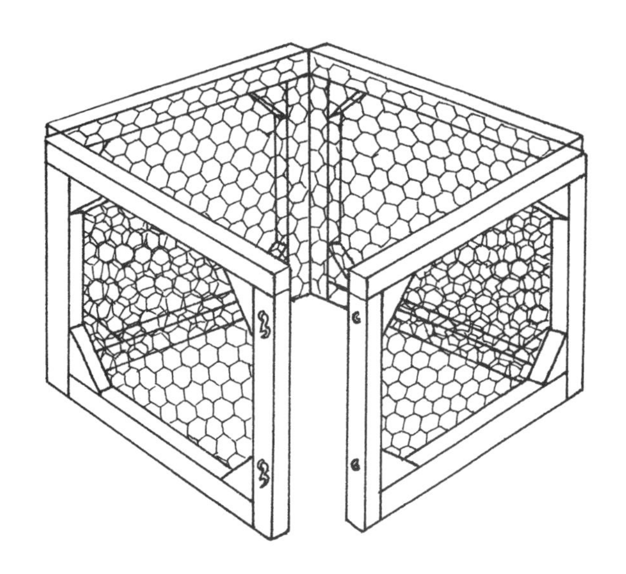

Wire mesh bin, wire-and-wood bin

- Wire mesh bins (see page 69) are simple and inexpensive to build. Wire-and-wood bins are more costly and more complicated. For stability and longer life, use ½-inch mesh hardware cloth; 1-inch chicken wire is less expensive and easier to work with but less stable and doesn't last as long. A good compromise is ½-inch chicken wire. You can purchase kits (usually sturdy, plastic-coated wire panels) that are even easier to assemble but slightly more expensive.
- Easy to turn and aerate because bin can be removed from around pile. To turn, place empty bin next to pile and fork material back into bin, mixing as you go. Almost effortless to move when empty.
- Mostly dog-proof, but not raccoon- or rodent-proof (except for kits with lids), so may not be suitable for urban areas. Does not camouflage compost pile.

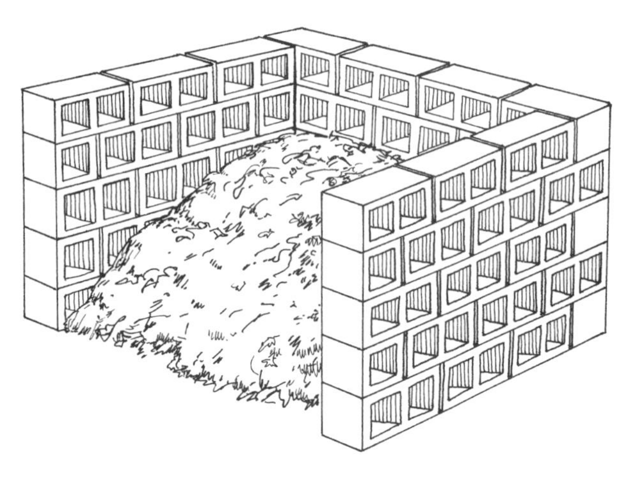

Cement block bin

- Simple to build (requires no special skills); somewhat more expensive than wire bin. Area must be level. A single, 3-foot by 3¾-foot bin requires 43 standard cement blocks (7½ x 7½ x 15 inches), plus 4 half blocks. Lay a row of blocks three deep on the sides and four deep across the back, facing holes outward for aeration. Stagger joints for stability, using half blocks at the front edge of alternate rows. Five rows will give you a 3-foot-tall bin.
- Gives easy access for turning and aerating. Must reconstruct to relocate.
- Provides easy access to animals. Does not camouflage compost pile unless sited so that only the cement blocks are visible.

Straw bale bin

- Simple and inexpensive to build. Stack straw bales three high to form three sides of a square. Reuse until bales start to break down (then pull apart and compost them, buying new ones for the sides). Additional bales (or a thick layer of straw) on top provide insulation to extend the composting season in cold climates.
- Offers easy access for turning and aerating. Must reconstruct to relocate.
- Provides easy access to animals, so not suitable for urban areas. Doesn't camouflage compost pile unless sited so that only the bales are visible.

Wooden bin

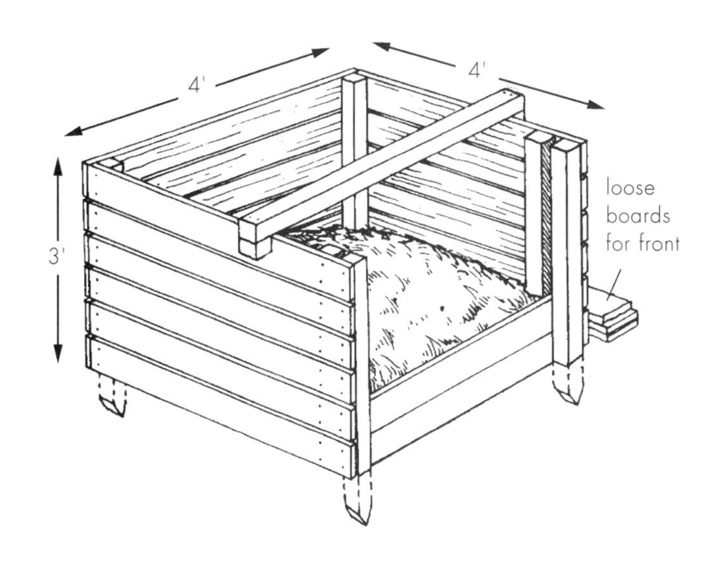

- Requires time and woodworking skills to build. Expensive, especially if you use rot-resistant cedar rather than pressure-treated wood. Reduce cost and improve aeration by using hardware cloth for the sides. Front boards should be easy to remove. Kits are easy to assemble but expensive; there are kits for multiple bins as well as single bins. Sturdy and long-lasting.
- Hard to turn and aerate, unless multiple bins are built side-by-side. Hard to move or relocate, so choose a good site.
- Animal-proof if covered and closed in front. Attractive.

Tumbler

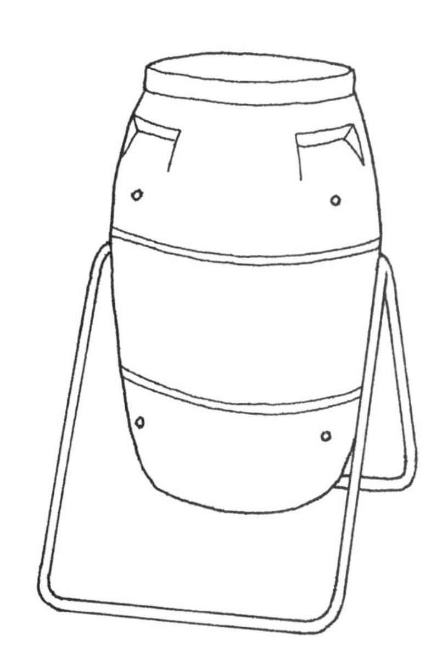

- Expensive, but requires little time or effort to assemble.
- Easy to turn and aerate. Usually doesn't hold large enough volume for hot composting, but always relatively fast for cold composting if material is chopped, loaded all at once, and turned frequently. Types that turn end-over-end or spheres that roll do a better job of mixing than cylinders that spin horizontally. Easy to move when empty.
- Animal-proof, so good for urban areas. Attractive.

Stationary plastic bin

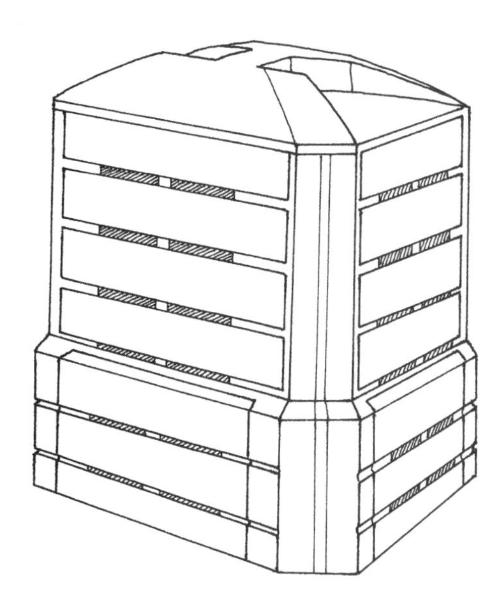

- Expensive, but requires little time or effort to assemble. Available in a range of sizes and prices. Many are now made from recycled plastic. See "A Garbage-Can Composter" on the next page for an inexpensive alternative.
- Hard to turn and aerate, except for models that come apart and reassemble easily, so best for cold composting. (Smaller models don't hold enough for hot composting, anyway.) Decomposition is slow unless material is chopped, added all at once, and aerated frequently. Easy to move when empty.
- Lids make these animal-proof. Attractive.

A Garbage-Can Composter

A plastic garbage can makes a tidy and inexpensive bin that fits in even the smallest yards. It's perfect if you have only small amounts of material to compost, such as kitchen scraps plus a few garden trimmings.

Once your bin is full, let it sit at least a couple of months so that everything inside can break down. (It will take longer if you don't stir or aerate the contents.) If you're in a rush to use compost, you can sift or dig out the older, finished material in the bottom and toss the rest back in the can. For nonstop composting, start a second bin and move the first to an out-of-the-way place until its contents are ready. *(For metric equivalents, see "Useful Conversions" on page 208.)*

1 Drill 8 to 10 holes in bottom of can. Space holes about 4 or 5 inches apart up the sides (stay at least 4 inches away from handles and rim). Don't drill holes in lid.

MATERIALS

- Heavy-duty, 8- or 10-gallon plastic garbage can with lid
- Drill with ½-in. bit
- Rubber tie-down strap (to hold lid securely in place)
- 3 bricks
- Tray (optional)
- Covered bucket or small garbage can filled with dry (loose) soil or sawdust

2 Pick a convenient spot. Arrange the three bricks so that they provide a stable base. If your can sits on a patio or deck, place a tray underneath the bricks to catch any drips (they can stain). Place 2 to 4 inches of loose, dry material (chopped leaves or straw) in the bottom.

3 Cover every additional 4-inch layer with an inch or so of soil or sawdust (or 2 inches of chopped leaves). Or sprinkle the soil or sawdust over every addition to soak up excess moisture and keep away flies. Attach the lid securely with the tie-down strap to keep animals out. Try rolling the can around periodically to mix and aerate the contents. Add enough water to moisten ingredients if they dry out.

A Simple Large Bin

If you want a sturdy, simple-to-make bin that can handle larger volumes of material, wire mesh is the answer. A five-panel bin is more effective than a square one, as the material in the corners of a square bin never heats up and takes a long time to break down. A cylinder of 36-inch-wide chicken wire is even simpler but flimsier. *(For metric equivalents, see "Useful Conversions" on page 208.)*

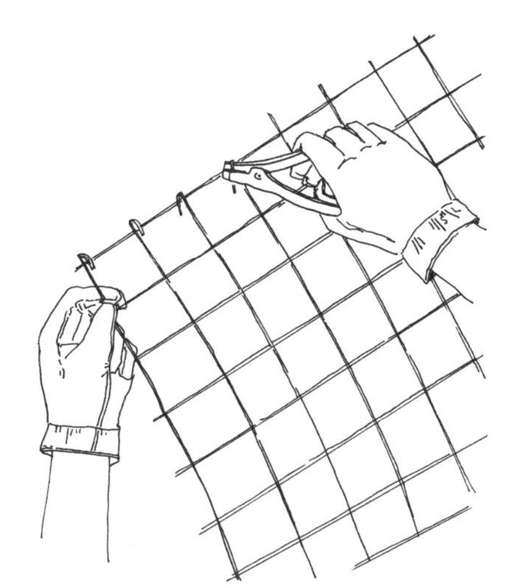

1 Cut five 3-foot-long sections of wire mesh. Leave long wires along one cut edge of each panel (the top), and cut those on the other edge (the bottom) flush with the horizontal wire. Using pliers, bend over and clamp each long wire flush with the mesh.

MATERIALS

- 15 feet of sturdy, 2-ft.-wide plastic-coated wire mesh (12- or 16-gauge)
- 20 metal or sturdy plastic clips (or plastic-coated flexible wire ties)
- Heavy-duty wire or tin snips
- Pliers
- Work gloves

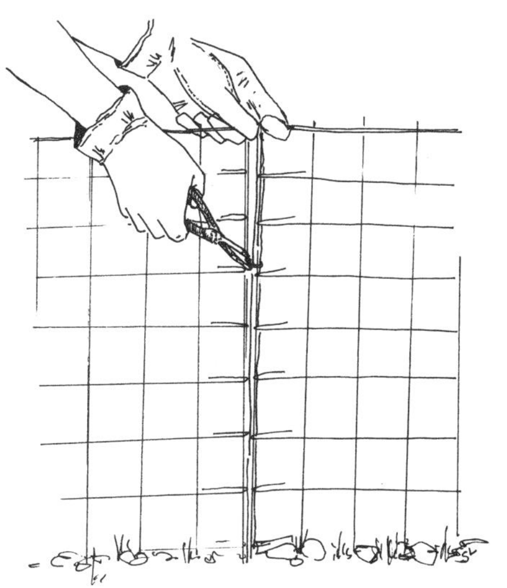

2 Attach the long edges of each panel to the adjacent panel, forming a pentagon. Clip or tie each "seam" at the top and bottom, and in two places in between.

3 Find or create a level site large enough to hold two bins side-by-side. To turn and aerate, tug the bin up and off the pile. Place the empty bin next to the pile. Fork the piled material into the empty bin, mixing well.

Triple Bin

MATERIALS

- Four 10-ft., pressure-treated 2 x 4s
- Eight 6-ft., pressure-treated 2 x 4s
- One 9-ft., two 6-ft., and six 3-ft. 2 x 2s (pressure-treated)
- Two 6-ft. cedar 2 x 6s
- Nine 6-ft. cedar 1 x 6s
- 22 feet H-in., 3-ft.-wide hardware cloth (to be cut into one 9 ft. piece and four 37-in. pieces)
- Twelve H-in. carriage bolts, 4 in. long, with washers and nuts
- 3 lbs. 16d galvanized nails
- H lb. 8d galvanized casement nails
- 250 chicken wire staples
- Two 10-ft. sheets clear corrugated fiberglass roofing (4 oz.; 2 feet wide)
- Three 8-ft. corrugated wood strips
- 40 gasketed aluminum nails for corrugated fiberglass roofing
- Two 3-in. zinc-plated hinges for lid
- Four flat 4-in. corner braces with screws
- Four flat 3-in. T-braces with screws

TOOLS

- Hand saw or circular power saw
- Drill with H-in. bits (for bolts) and J-in. bits (for nails)
- Screwdriver
- Hammer
- Tin snips or sturdy wire cutters
- Tape measure
- Pencil
- I-in. socket wrench
- Carpenter's square
- Power stapler (optional)

Bin design modified from original plan developed by Seattle Tilth Association for Seattle's Master Composting Program. Reprinted with permission from Seattle Tilth and the Seattle Engineering Department.

For composting large volumes, nothing beats the triple-bin system. The first bin holds new, accumulating layers. When this bin is full, its contents are turned into the second bin for curing. Once most of this material has broken down, the contents of the second bin are forked into the third bin, which holds compost until it is finished.

These directions are for the smallest dimensions that allow hot piles. For hot (fast) compost, chop all materials and turn the piles in place every week to 10 days, in addition to turning them into the next bin. This design uses wire mesh for maximum aeration and to reduce cost. You can substitute cedar boards or pressure-treated wood for the sides; space them at least H inch apart for air circulation. Wear work gloves when handling hardware cloth, and wear eye and ear protection when cutting lumber and assembling bins. *(For metric equivalents, see "Useful Conversions" on page 208.)*

Build Dividers

1 Cut one 31½-inch and one 36-inch piece from each 6-foot 2 x 4. Butt-end nail the four pieces into a 35 x 36 inch square. Repeat for other three sections. Cut four 37-inch-long sections of hardware cloth; bend edges back 1 inch. Stretch cloth across each frame, check frame for squareness, and staple screen tightly into place every 4 inches around edge.

2 Cut 9-foot pieces out of the four 10-foot boards; two of these will be baseboards and one will be the top board at the back (save the last board for step 6). Position the four dividers on edge parallel to one another, 3 feet apart. Measure and mark center points on the two inside dividers. Position the two baseboards on top of dividers. Make outer dividers/sides flush with the ends of the baseboards. Measure 3 feet in from each end and mark positions for the two inside dividers. Line up the center lines with the 3-foot marks, making front edge flush with the baseboard. Drill a ½-inch hole through each junction, centered 1 inch from the inside edge. Secure baseboards with carriage bolts, but don't tighten yet.

3 Turn the unit right-side up and repeat the process for the top 9-foot board in back. Using the carpenter's square (or by measuring between opposing corners and adjusting until measurements are exactly equal), make sure bin is square; tighten all bolts securely. Fasten a 9-foot-long piece of hardware cloth securely to back side of bin, driving staples every 4 inches around the frame.

Make Front Slats and Runners

4 Cut four 36-inch-long 2 x 6s for front slat runners. Nail one securely to the front of each outside divider and baseboard, making it flush on top and outside edges. Center the remaining boards on the front of the inside dividers (flush with top edge and overlapping about 1 inch on each side); nail securely in place.

5 Cut the six 3-foot 2 x 2s down to 34 inches long for the inside (back) runners. Nail one inside runner parallel to the front runners on the side of each divider, leaving a 1-inch gap for slats. Cut all the 1 x 6 cedar boards into slats 31 1/4 inches long and insert in runner slots. Do not nail in place; you'll want to remove these slats to get at your compost. The runners allow you to easily slide the slats in and out.

Make Fiberglass Lid

6 Use the last 9-foot 2 x 4 for the back of the lid and the 9-foot 2 x 2 as the front. Cut the 6-foot 2 x 2s into four 32 1/2-inch lengths for sides and center braces. Lay out into position on ground and check for squareness. Screw in corner braces and T-braces on bottom side of frame. Center lid frame (brace-side down) on bin structure and attach with hinges. Cut corrugated wood strips to fit the front and back 9-foot sections of the lid frame. Predrill nail holes in the strips with 1/8-inch drill bit; nail with 8d casement nails. Cut fiberglass into approximately 3-foot lengths to fit flush with front and back edges. (Ridges run back to front.) Overlay pieces at least one channel wide to keep out rain. Predrill each nail hole in fiberglass and corrugated wood strips on top of every third hump; nail with gasketed nails.

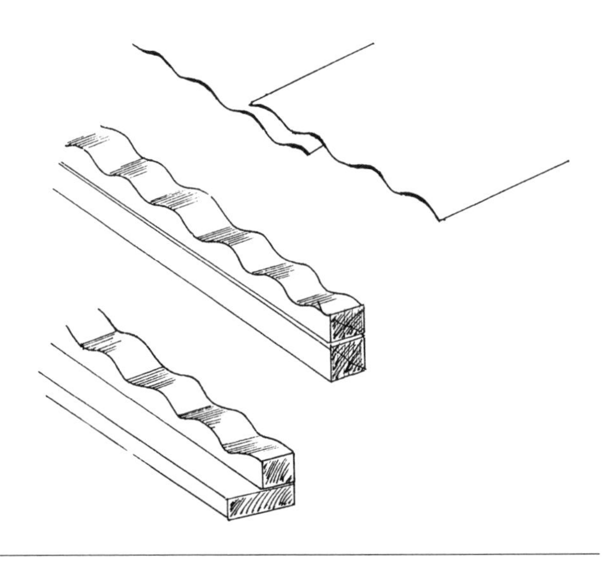

Composting with Earthworms

If you want to continue composting through cold winters or blistering summers, indoor earthworm bins are the answer. They also work well for apartment dwellers who want to recycle kitchen scraps, and for people with tiny yards. Worm castings are richer than other forms of compost; they're one of the best soil conditioners and one of the most balanced sources of nutrients. You can buy bins complete with worms, but it's very easy to make your own (and less expensive). *(For metric equivalents, see "Useful Conversions" on page 208.)*

- **Locate your bin in a basement,** storage room, or garage that stays above 40° (5°C) and below 90°F (32°C). Try to keep bins at 60 to 70°F (15 to 22°C), where composting worms are most active.
- **Order red wigglers or red worms** *(Lumbricus rubellus)* or brandling worms *(Eisenia foetida);* suppliers are listed in the appendix. If you know someone who composts with worms, ask her for 3 or 4 cups of bedding with lots of worms. Ordinary earthworms won't work.

Materials

- Plastic bin with handles and matching lid, approximately 1½ x 2 feet, and at least 8 in. tall
- Drill with ½-in. bit
- Plastic tray as large as bin (or sheet of rigid plastic large enough to cover top)
- 2 pieces scrap 1 x 2 or 2 x 4 wood, about 1 ft. long
- Fiberglass window screening, 24 x 18 in.
- Newspaper or chopped leaves
- Spray mister or watering can with sprinkler nozzle
- Sturdy, long-handled plastic spoon
- About 2 lbs. red wiggler worms

1 Drill 12 holes in bottom of bin, 4 to 5 inches apart. Or melt holes with a fat (size 15) metal knitting needle heated in a flame; use a potholder and open a window to vent fumes.

2 Lay wood scraps flat on tray, and place bin on top. If you can't find a large tray, use the lid as your tray and make a new lid out of a sheet of rigid plastic. (A plastic garbage bag works if you punch a few small holes for air.)

3 Lay fiberglass screening over bottom of bin; it should come up an inch or two on the sides. Fill bin halfway with moistened bedding. Rip newspapers into 1- to 2-inch-wide strips and mist or sprinkle until moist but not soggy. Or moisten chopped leaves for bedding. Release worms on top of bedding.

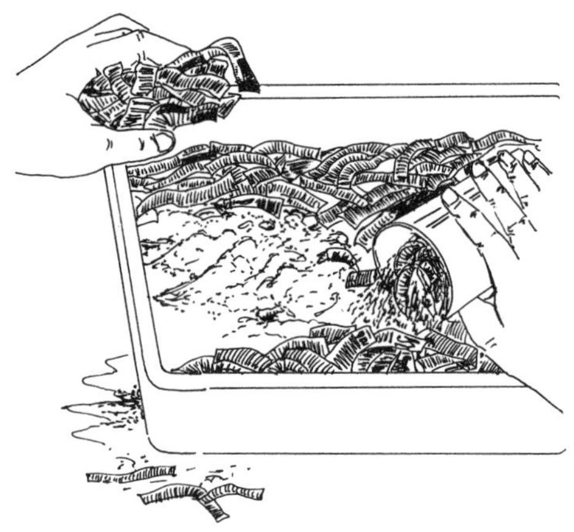

4 Feed your worms vegetable (or garden) trimmings and/or coffee grounds. Don't add anything from the "What to Avoid" list on page 57, and don't add eggshells (worms avoid them). Peel back bedding, add the scraps, and replace bedding to minimize odors and fruit flies.

5 Start with a couple of pounds of food a week. After several feedings, use the spoon to dig a hole for new additions. Or spread additions on top of existing bedding and cover with fresh, moistened bedding (or a sprinkling of sawdust). If you use a lot of acidic material (coffee grounds and sawdust), sprinkle a tablespoon of lime over the bin every few months.

6 Leave the lid ajar for ventilation. If the contents start getting soggy, pull the lid back to let in more air. If they start to dry, close the lid or mist with water. You can close the lid completely for a week or two at a time without harming the worms. Keep bedding at least 3 inches below rim, and loosen periodically with spoon.

7 Whenever most of the material looks like soil, you can remove castings by scraping finished compost off the top, a little at a time. Or push all the compost and worms to one side. Fill the empty half with new, moistened bedding. Add food scraps only to the new bedding. After a few weeks, many worms will have moved into the new half, so you can dig out the old and leave most worms behind. Repeat whenever one side appears ready for garden use.

Outdoor Worm Bins

Here's an easy variation on earthworm composting. In summer — or where winters are mild — try outdoor worm bins. Start with an 8- or 10-gallon plastic trash can. Cut out the bottom and drill or punch several holes in the sides. Dig a 1- to 2-foot hole and sink the can partway in the ground. Fill as for indoor bins and replace the lid. Check moisture level often and add dry bedding if soggy or sprinkle with water if dry.

The worms will slow down in cool weather and die if it gets cold. Try covering with hay bales for insulation, but an indoor bin to overwinter the worms is a better idea. If you dig up earthworms for outdoor bins instead of using red wigglers, they'll simply retreat deep into the soil to overwinter, returning the following spring.

MASTER GARDENING TIP

How Hot Is Hot?

To find out how hot your pile gets, you'll have to take its temperature. Specialized compost or soil thermometers are sturdy enough to push into piles. Well-made hot piles usually reach about 160°F (71°C). To kill most weed seeds and diseases, your pile needs to maintain this temperature for a few days. Only the most heat-resistant weed seeds and some viruses will then survive. If a pile gets to only 120°F (49°C), it may take a couple of months to kill the same organisms, and some will probably survive. If you don't want to worry about temperature, simply bury pest- or disease-ridden plants and troublesome weeds away from the garden rather than composting them.

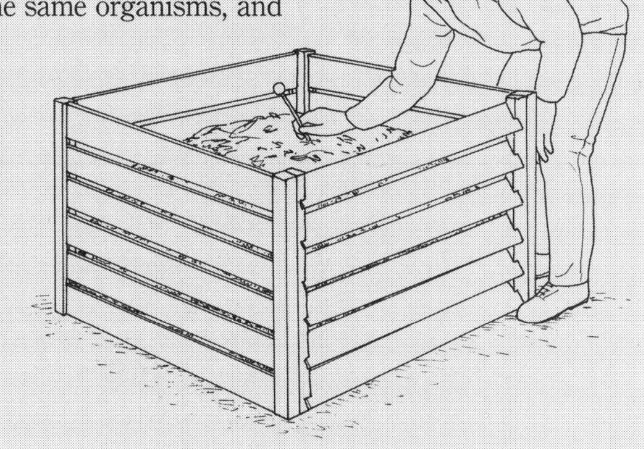

HINTS FOR SUCCESS

Dealing with Fruit Flies

- Fruit flies can sometimes be a problem around compost buckets or indoor earthworm bins. You can minimize fruit flies by keeping buckets covered and by carefully covering each addition of food (or at least fruit) scraps to your bin. To minimize the risk, avoid storing uncovered fruit in the same room.
- Vacuuming is one of the easiest ways to control fruit flies. Turn on the vacuum, open the lid, and whoosh! watch them disappear. Empty the vacuum bag outdoors. Hanging old-fashioned, sticky flypaper strips above bins works for small numbers of flies.
- Disposable, baited traps are now available from suppliers that sell organic controls. To make your own, make a paper funnel with an opening no larger than ¼ inch (.6 cm). Pour ¼ cup (60 ml) of beer into a jar. Set the funnel on top. Run tape around the top edge of the jar to eliminate any gaps between the jar and the funnel. Fruit flies will fly in and drown in the beer. Beer works better than fruit baits because the flies can't lay eggs in it. If you don't want to use beer, try molasses (it's harder to wash out of the jar, though).

Sheet Composting

Sheet composting just means working horizontally, spreading material in wide layers (sheets) over the ground. It's very simple; you let earthworms and soil microbes do the composting and mixing. No bins, no turning. Any nutrients that leach will be absorbed by the soil below for future plant use. It's a great way to start a new bed, especially in a lawn — you skip the hard work of digging and removing sod. It's also great for lucky people who have more material to compost than they can keep up with. All it takes is planning ahead.

Start your sheet compost six months before you want to plant. For a spring garden, start in fall. If you start in spring you can plant by late summer (compost works faster in warmer weather). Use the same ingredients as for any other type of compost. With heavy soil, you may eventually wish to double-dig (see page 141) or build raised beds (see page 139). Either will be much easier after six months of sheet composting loosens your soil! *(For metric equivalents, see "Useful Conversions" on page 208.)*

MATERIALS

- Lawn mower or string trimmer
- Pruning shears or saw (for woody plants)
- Lots of newspaper or corrugated cardboard
- Limestone or sulfur
- Ordinary compost materials (about 1½ cubic yards/50 sq. feet)
- Wood chips or other attractive mulch for top layer

1 Mow existing plants as short as possible, leaving clippings. Cut off any woody plants at ground level and chop into small pieces. If clippings are sparse, sprinkle a dusting of another nitrogen source over area (manure, alfalfa meal, or fresh grass clippings). If your soil's usually acidic, sprinkle some limestone; substitute sulfur for very alkaline soils. Sprinkle some granite meal to ensure a micronutrient supply if you wish.

2 Spread a weed-smothering layer of newspapers (8 to 10 sheets thick) or cardboard over area. Overlap edges by 4 inches so no weeds can poke through. Dampen newspapers to make them easier to work with.

3 Spread a 3-inch layer of manure or compost. You can substitute chopped leaves, but try to include a little compost or manure. You'll need about half a cubic yard to cover 50 square feet.

4 Add a 6-inch layer of coarse material such as garden trimmings, wood shavings, straw, kitchen scraps, and chopped leaves. Try to use a mixture. Mix grass clippings or sawdust with other materials — or with each other — to keep either from forming a waterproof layer. You'll need a cubic yard to cover every 50 square feet. Try for 9 to 12 inches of total material on top of the smothering layer. Water well to soak all materials.

5 Spread more newspaper (six to eight sheets thick) or cardboard, overlapping edges, to smother any weeds or seeds in the preceding layer. Cover with 2 to 3 inches of wood chips, pine needles, or other attractive material to improve appearance and hold newspaper in place. Sprinkle water over this layer to help hold it in place. Now just sit back and wait six months.

HINT FOR SUCCESS

Sheet Composting in Existing Gardens

Many people are already sheet composting but don't realize it. Spreading a thick layer of chopped leaves, used straw mulch, or similar compostable material over your garden and turning it under is another form of sheet composting. You can rototill the material or turn it under by hand with a garden fork or shovel.

If you do this in fall, either leave the soil in large clods or spread a thin layer of chopped leaves over the soil to minimize erosion and leaching. The leaves will be converted into humus by spring. You don't even have to turn material under, if you don't mind waiting longer for it to break down.

Pit or Trench Composting

The simplest way to compost, if you don't mind digging, is right in the soil. Dig a hole or trench in the ground, fill it with kitchen and/or garden trimmings, and cover with the removed soil. In a few months, even the most stubborn clay will be crumbly, easy to dig, and loaded with earthworms. Pit and trench composting don't give you compost to spread, but they work wonders on the soil in that spot.

Trench composting is perfect if you want to return kitchen and/or garden wastes to the soil but don't want to worry about carbon/nitrogen balance or turning piles. It's an excellent way to improve the soil for a new bed if you can plan six months ahead. As with sheet composting, if you eventually decide to double-dig or build raised beds, you'll have made the soil much easier to dig and improved its structure.

Pit composting is a good way to dispose of diseased plants. Use it in addition to a regular pile, if you don't want to worry about whether your pile gets hot enough to kill diseases. Just locate your pit away from the garden and cover diseased material immediately with several inches of soil. Here are several easy variations.

MASTER GARDENING TIPS

Words to the Wise

- Don't try to dig your trench when the soil is bone dry. It's hard work. Wait until soil is moist (spring or fall in most areas).
- Don't dig your trench when the soil is wet. Use the squeeze test (page 10) to see if your soil is ready to dig.
- Where the soil freezes, you'll need a waterproof container filled with dry sand or sawdust to cover kitchen scraps during winter. Spread soil over top in spring.

For a new small bed, make your trench the size of the desired bed. For beds larger than 2 feet by 4 feet, dig only about that size at one time; after filling the first trench, repeat until you've covered the entire area.

Remove sod and set aside. Dig out soil to about 1½ feet deep. Lay removed sod upside down in the bottom of the hole. Add plant debris, covering every new addition with some of the removed soil. If you add 4 to 6 inches all at once, cover this thick layer with a couple of inches of soil. Once the trench is filled, spread any remaining soil over the top. Let it sit six months and you're ready to plant.

If you want to improve the soil in an area of your garden, plan on keeping it out of cultivation for a season. Dig one or a series of trenches, as described on page 78. Fill with any of the usual compost materials and let sit until turning over a shovelful reveals finished compost. You'll notice a tremendous difference the following growing season.

Year one. Dig trenches next to your garden beds, leaving a walking space. Simply pitch garden debris into the trenches, adding a thin layer of straw, dried grass clippings, or chopped leaves from time to time. Be sure to remove any weed seeds and infested or diseased plants first (bury them well away from your garden). Fill in the compost trenches with soil and organic matter and leave in place over the winter.

Year two. Dig a new trench where your garden bed was the previous season. Allow the organic matter in the old trench to decompose completely by using that space for a walking space.

Year three. Plant your new bed in the humus-rich soil of year one's compost trench. Fill in year two's compost trench for a walking space and begin composting in the old garden bed. You've now completed a full cycle and can begin again from the top in the next season.

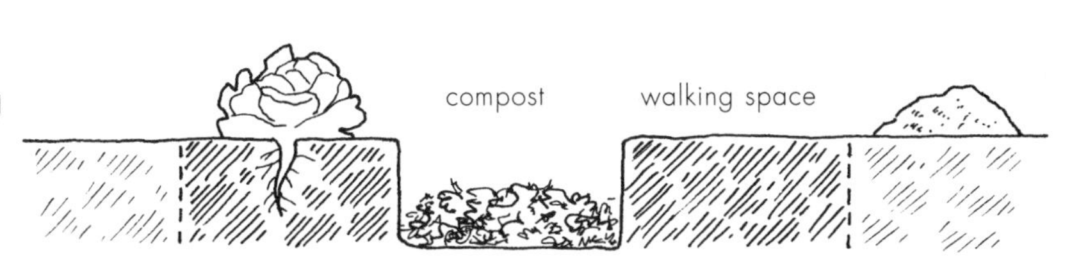

Using Your Compost

Compost is superb as a soil-improving amendment, topdressing, and mulch. Finished compost can be used anywhere in the garden. To maximize its benefits, use finished compost within a few months. Like well-aged manure, though, even the oldest compost still improves soil.

Gardeners rarely have enough compost. Ideally, all beds (vegetable, perennial, shrub, even lawns and trees) should get an inch of compost every year. An inch is enough to provide fresh cultures of soil organisms and to maintain soil that's already in good shape. If you're trying to correct physical problems such as poor drainage or structure, or if your soil's very low in organic matter, you'll need to use even more. In warm, humid climates where organic matter breaks down very rapidly, try for 2 inches.

If you don't have enough compost to go around, concentrate on trouble spots — soil that has poor structure or drainage or is lacking in nutrients, and hot spots where diseases or pests are a problem. You can get away with half as much if you use green manures or maintain a constant cover of organic mulch.

DID YOU KNOW?

A Midwestern study compared compost with stockpiled feedlot manure. Test plots using only compost produced yields similar to those fertilized with four times the volume of manure! The compost raised pH and levels of organic matter, nitrogen, phosphorus, and potassium more than could be accounted for by the nutrient content of the compost. The study concluded that compost must increase the availability of existing soil nutrients while also supplying small additional amounts.

Add a handful of finished compost to the bottom of each hole when transplanting small seedlings. Add a shovelful for larger plants and heavy feeders such as tomatoes, melons, and squash. Sprinkle finished compost into rows when planting seeds.

For use in potting soil, or to spread on (topdress) lawns in spring or fall, screen finished compost to remove large pieces. For most uses, ½-inch mesh is fine enough. Staple hardware cloth to a frame of 2 x 2 or 2 x 4 wood. If you add legs to one side of a large screen, you can toss shovelfuls against the screen to sift (either onto the ground or into a garden cart). Or size the frame to lay across the top of a cart or plastic garbage can so that finished compost falls into the container.

You can spread unfinished compost over the garden in fall. It will finish breaking down by spring and may be left on the surface or tilled under. Avoid spreading unfinished compost around vegetables and flowers as a topdressing. To use it as mulch on annual flowers or vegetables, spread a layer of finished compost underneath. You can use unfinished compost alone as a mulch for shrubs and trees.

Finished, screened compost is an excellent ingredient in soil mixes for houseplants and even for starting seedlings. (You don't have to sterilize it; unsterilized compost contains organisms that suppress ever-present damping-off fungi.) If the compost is newly made, use the test on page 61 to make sure it's really finished. You may want to test commercial compost, too. It isn't necessarily finished by the time it reaches you. Unfinished compost can harm the germination and growth of many vegetable and flower seedlings.

Compost can be used as a foliar fertilizer in the form of compost tea (see page 129). This liquid form of compost can be sprayed on or poured over leaves, or used to water houseplants. Compost tea is great for giving plants a nutrient boost and controlling some diseases.

Compost Troubleshooter

You can solve most composting "problems" simply by waiting. Most material will break down eventually, without any effort on your part. If you don't want to wait a year or more for nature to take its course, follow the remedies below.

Symptom	Cause	Remedy
Pile doesn't heat up, feels dry.	Too dry	Add water until materials in center feel evenly moist; in dry climates, cover pile with tarp and water whenever it dries out.
Pile doesn't heat up (or heats up only in center); feels moist.	Not large enough	Make sure pile is 3 feet square (90 cm^2) and at least 4 feet (1.2 m) tall, or wait for pile to undergo cold composting.
Large pile doesn't heat up, feels moist.	Not enough nitrogen	Add alfalfa meal, manure, fresh grass clippings, or other nitrogen (protein) source and turn pile.
Pile cools off before most material has decomposed.	Needs to be turned	Turn pile with garden fork, mixing material in center with outer or undecomposed material.
Pile smells bad, feels soggy or wet.	Too wet	Add shredded newspaper, straw, or other dry, carbon-rich material and turn pile; in wet climates, cover pile to keep off rain.
Pile smells bad (like ammonia), feels moist, not soggy.	Too much nitrogen or not enough air	Add shredded newspaper, straw, or other carbon-rich material (plus water to dampen) and turn pile to aerate; use less nitrogen-rich material in future piles and turn more often.
All material doesn't break down.	Too dry or not enough nitrogen	For soil that is too dry, add water until materials in center feel evenly moist; in dry climates, cover pile with tarp and water whenever it dries out. For soils needing nitrogen, add alfalfa meal, manure, fresh grass clippings, or other nitrogen (protein) source and turn pile.
Matted layer doesn't break down.	Needs mixing	Turn pile, breaking up matted layer and mixing with other material.
Some pieces didn't break down.	Pieces too large, too woody, or not biodegradable	Sift compost; in future, leave out whatever didn't break down, or chop into smaller pieces and add more nitrogen-rich material.

CHAPTER·5

Mulches, Amendments, and Green Manures

Plants and soil organisms like a varied diet. Adding organic matter in different forms — mulches, topdressings, soil amendments, and green manures — ensures that both get a balanced supply of micronutrients. It can also make life easier for the gardener, as different methods of adding organic matter suit individual budgets, schedules, and energy levels.

Some methods are better suited to specific situations. Green manures are great when starting new gardens, and in a crop rotation in vegetable or annual beds. Maintaining a 2-inch-thick layer of coarse compost mulch keeps an ornamental shrub border, flower bed, or foundation planting supplied with organic matter. If slugs or snails keep you from using mulch, substitute a short-term, well-decomposed topdressing. When you renovate a perennial bed or divide the plants in a large area, incorporate soil amendments to give plants a head start, supplementing each year with topdressing or mulch. Whichever method you choose, use overall garden health and performance, plus periodic earthworm counts, as indicators of adequate organic matter.

Striving for high yields in the vegetable garden is one great way of building organic matter. The same soil conditions that promote the highest crop yields also promote a large and healthy community of soil organisms: ample amounts of organic matter, slightly acidic pH (around 6.5), and a balance of all essential nutrients. Plants that produce high yields have extensive root systems, which are left in the soil after harvest. Such plants also bear fruit, but produce lots of leaves and stems. This means more plant residues to turn under at the end of the season, or to add to the compost pile.

In This Chapter

Mulches

Mulching is one of the easiest ways to add organic matter: Spread the material on top of the soil and let earthworms do the tilling for you. Mulching (or its close cousin, topdressing) is often the only way to supply organic matter to trees, shrubs, berry bushes, and perennials. If these are shallow-rooted, they'll resent any attempt to dig in amendments. It's much safer (and simpler) to spread organic materials on top of their roots. As a bonus, mulching greatly reduces watering and weeding chores. In addition to adding humus, mulches improve the soil in many ways, from reducing erosion to encouraging soil organisms.

Organic Mulches

Not all mulches provide organic matter and nutrients — only organic ones do. Organic mulches can be anything you'd put in the compost heap. For mulching, though — unlike composting — the longer something takes to break down, the better. Wood chips and bark nuggets that are too hard for the compost heap are perfect for mulch. They don't need to be replenished as often as a mulch of grass clippings.

A few warnings are in order before you mulch everything in sight. In wet soils and very humid climates, organic mulches can promote some diseases and increase chances of rot. If spread too thickly, they can interfere with the soil's air circulation. Fresh, unweathered carbon-rich mulches such as sawdust can temporarily tie up enough soil nitrogen to interfere with plant growth. This problem is easily solved by letting them weather until their color fades, or by spreading nitrogen-rich material underneath them. Often, these same materials are flammable and should be used with care in fire-prone areas.

Inorganic Mulches

Inorganic mulches provide some benefits, although they don't supply organic matter. It's a good idea to spread 1 to 2 inches of compost or aged manure over the soil before using inorganic mulches. Once they're in place, it's hard to replenish soil organic matter. Here are a few inorganic mulches, and reasons why you might want to consider using them.

- **Black paper and black plastic.** These are excellent for warming soil in spring to lengthen the growing season in cold climates; also good for weed control.

- **Gravel.** A decorative mulch, gravel is good around plants that need excellent drainage or are prone to rotting (most alpine, desert, and rock-garden plants); it also keeps soil cool in hot climates, prevents wind erosion, and is effective where slugs or snails are a major problem.

DID YOU KNOW?

Why Use Mulch?

- ▶ Keeps soil moist.
- ▶ Smothers most weeds.
- ▶ Keeps soil soft so that any weeds are easy to pull and more rain soaks in.
- ▶ Shades soil, reducing heat stress for plants and soil organisms.
- ▶ Keeps soil warmer on cold fall nights.
- ▶ Prevents sudden swings in soil temperature, increasing winter survival of plants.
- ▶ Helps prevent plants from heaving out of soil in winter.
- ▶ Reduces wind and water erosion.
- ▶ Prevents puddling or crusting from hard rains.
- ▶ Reduces or prevents soilborne diseases by keeping rain from splashing soil onto plants.
- ▶ Keeps fruits and vegetables clean.
- ▶ Minimizes stress for new transplants.

If the mulch is organic, it also:

- ▶ Adds organic matter and humus.
- ▶ Feeds and encourages earthworms and other soil organisms.
- ▶ Conditions soil.
- ▶ Slowly adds nutrients to soil.
- ▶ Increases retention and availability of nutrients.
- ▶ Recycles lawn, garden, and tree trimmings.

• **Landscape fabric.** These are plastic materials with fine holes that let air and moisture through but not weeds. They're good for weed suppression around shrubs if covered with a decorative mulch and particularly useful for preventing weeds in gravel or stone paths and patios.

Topdressings

Mulches that disappear very quickly are more properly called topdressings. The purpose of a topdressing is to get organic matter and nutrients into the soil as quickly as possible. Mulches need to stick around a little longer to fulfill their main purpose, which is providing a protective blanket for the soil. Providing humus and nutrients is really a secondary benefit of mulches, especially since they often take a long time to break down and release either.

HINT FOR SUCCESS

Bury Your Peat Moss

Peat moss makes a good soil amendment but a bad mulch. It forms a water-repellent crust that sheds instead of absorbs rain. On top of the soil, it takes forever to break down, so it's even a poor source of organic matter. It's also a poor source of nutrients.

Don't use peat moss as mulch — there are much better options! You can buy easy-to-spread, bagged compost just about anywhere you'd buy peat moss. Save peat moss for digging into soil, and then only for acid-loving plants.

MASTER GARDENING TIPS

Seasonal Tips for Mulching

- ***Spring.*** In cool climates, rake mulch off vegetable and annual beds a few weeks before planting time to let soil warm. Once danger of hard frost has passed, pull mulch away from perennial plants. When plants put out new growth, soil is warm enough to replace mulch. Wait until soil is warm before mulching heat-loving annuals and vegetables such as tomatoes and peppers.
- ***Summer.*** In hot climates, choose light-colored mulches to reflect heat and keep soil cool. In climates that stay cool all summer, choose dark-colored mulches to absorb heat (consider paper and plastic mulches). In warm, humid climates, organic mulches break down very quickly; be prepared to replenish frequently.
- ***Fall.*** In vegetable gardens, leave soil bare for a few weeks before spreading winter mulch. This gives birds time to dig through the soil and eat pests and pest eggs. In late fall, cover empty vegetable and annual beds with chopped leaves or other mulch to reduce soil erosion and nutrient leaching.
- ***Winter.*** A thick layer of loose mulch insulates plants from the alternate freezing and thawing that can weaken or kill them, or even toss them out of the ground. Snow is one of the best insulators, but more predictable substitutes are needed. Chopped leaves, loose (weed-free) straw, pine needles, and evergreen branches (as from a discarded Christmas tree) work well. Wait until hard frost kills back plants and the top inch of soil freezes before piling on the mulch. In perennial gardens, leave a few stems on spent plants to help hold the mulch in place, but remove any diseased stems and leaves before mulching.

  Rodents love to nest in thick mulch and may gnaw the bark of thickly mulched trees and shrubs during winter. To protect trunks at ground level, wrap a 1-foot-wide strip of hardware cloth loosely around the trunk, overlapping the edges, and secure in place. Remove in spring.

Types of Mulches

Choose your mulches carefully for the best results. Consider how attractive you need them to be. Flower borders and plantings by the front door need a good-looking mulch; rows of raspberry bushes tucked out of sight don't. If cost is a major consideration, see what's available for free in your community. Some towns offer free wood chips, and many people are only too happy to give you their bagged leaves in the fall.

Material	Appearance	Insulation Value	Relative Cost	Thickness	Weed Control
Bark chunks and nuggets	excellent	good	high	3–4 in.	good
Bark, shredded	good	good	moderate	2–4 in.	excellent
Buckwheat hulls	good	good	high	1–1½ in.	good; may sprout
Burlap	poor	fair	moderate	1 layer	poor
Cocoa hulls	good to excellent	good	high in most areas	1 in.	good
Coffee grounds	good	fair	low or free	never more than 1 in.	good
Compost	fair	good	free or moderate	1–3 in.	fair
Corncobs, ground	good	good; heats up when wet	low in Midwest	2–3 in.	excellent
Cottonseed hulls	good	good	low in South	1–2 in.	good
Evergreen boughs	poor	good	low or free	1 or more layers	fair
Grass clippings	poor, unless dried	good	free	1 in. fresh, 2 in. dried	fair
Hay, dried	poor, unless chopped	good	moderate; low, if spoiled	6–8 in., 2–3 in. if chopped	good, unless it contains seeds
Hops, spent	fair	fair; heats up when wet	low, where available	1–3 in.	good
Landscape fabric (geotextiles)	poor, unless covered with other material	poor	high	1 layer	excellent
Leaves, chopped or shredded	fair	good	free	2–3 in.	good
Leaf mold	fair	good	low or free	1½ in.	fair to good
Manure, aged	fair	good	moderate	1–3 in.	fair

To cut back on work, choose a long-lasting mulch. In the vegetable garden, or to build up soil in a perennial bed, choose something finer or softer that breaks down more quickly. In a vegetable bed, finer or softer materials can be turned under the soil at the end of the season. This chart gives the characteristics of most common mulch materials. *(For metric equivalents, see "Useful Conversions" on page 208.)*

Water Penetration	Soil Moisture Retention	Speed of Decomposition	Comments
good	good	very slow	Best as long-term mulch around ornamentals; can be used over landscape fabric.
good	good	slow	Best as long-term mulch around ornamentals; more stable than chunks on slopes.
excellent	fair	slow	Easy to handle; lightweight; may be blown around in high wind or dislodged by rain.
excellent	fair	slow	Excellent for preventing erosion on slopes; new grass grows right through it; cut Xs to plant groundcovers or shrubs.
good, unless allowed to mat	good	slow	Sawdust can be added to improve texture and increase water retention; may develop mold (unsightly but not harmful); chocolatey smell; dark color absorbs heat.
fair; may cake	good	fairly rapid	Best used thinly or mixed with sawdust to avoid matting, which interferes with soil's air circulation.
good	good	rapid; supplies nutrients	Keep 1–2 in. away from plant stems; if using unfinished compost around annuals or vegetables, spread thin layer of finished compost or fresh grass clippings underneath.
fair	excellent	slow	Avoid close contact with plant stems because of heat generation; mix with nitrogen source to avoid depleting soil.
good	good	fairly rapid	Will blow in wind; has fertilizer value similar to cottonseed meal.
good	fair	slow	Good for wind protection and erosion control; excellent winter mulch; should be removed from perennials in spring.
good, unless matted	fair	rapid	Dried clippings are best; green clippings add nitrogen and are good under sawdust and other high-carbon mulches; fresh clippings can mat and smell bad unless spread thinly; compost first if herbicides were used on lawn.
good	good	rapid	Second- or third-growth hay that hasn't gone to seed is ideal; contains more nitrogen than straw but often more seeds, too.
good	good	slow	Avoid close contact with trunks and stems because of heat generation; supplies nitrogen and other nutrients.
fair	good	rapid, unless covered	Use of a cover mulch needed to increase longevity and improve appearance.
good	good	fairly slow	Chop with lawn mower or leaf shredder to reduce matting; contribute many valuable nutrients; chopped leaves good for winter protection; can be mixed with other mulch materials.
fair, if not thick	good	rapid	Use like compost; supplies similar nutrients.
fair to good	good	rapid	Should be at least partially rotted; supplies nitrogen and many other nutrients.

Material	Appearance	Insulation Value	Relative Cost	Thickness	Weed Control
Newspaper	poor, unless covered with other material	fair	free	several layers	good
Nutshells (pecan, walnut, almond)	excellent	good	low, where plentiful	1–2 in.	good
Oak leaves, chopped	good	good	low or free	2–4 in.	good
Oyster shells, ground	good	fair	high	1–2 in.	fair
Paper pulp	poor	fair	moderate	½ in.	fair
Peanut hulls	good	good	low, where plentiful	1–2 in.	good
Pea vines	good	fair	low or free	3–4 in., 2 in. if chopped	fair to good
Pine needles	excellent	good	low	1–2 in.	good
Planter's (black) paper	poor	fair; heats up soil	moderate	1 layer	good
Plastic	poor, unless covered	fair; heats up soil	low	1 layer (1–6 mil)	excellent
Salt hay	good	good	high, unless you gather it	3–6 in.	good; contains no seeds
Sawdust	fair to good	good	low	1–1½ in.	good
Seaweed	poor	good	low near coast	4–6 in.	excellent
Stone or gravel	excellent	good; dark colors heat up, light colors reflect heat	high	2–4 in.	fair, unless used over landscape fabric
Straw	fair, unless chopped	good	low to moderate	6–8 in., 2–3 in. if chopped	good, unless it contains many seeds
Sugarcane (bagasse)	poor to fair	good	moderate, where available	2–3 in.	good
Wood chips	good	good	moderate or free	2–4 in.	good
Wood shavings	fair	fair	low	2–3 in.	fair

Water Penetration	Soil Moisture Retention	Speed of Decomposition	Comments
fair	good	slow, if several layers are used	Overlap edges of sheets; works well if shredded; avoid glossy inserts with colored ink; flammable when dry.
good	good	very slow	Supplies micronutrients, but very slowly; resists fire.
good	good	slow	Recommended for acid-soil plants; less attractive to slugs and other pests than mixed leaves; chop before using; slightly acidic.
good	good	slow	Works like lime to raise soil pH; good mixed with other mulch materials.
fair	good	rapid	Requires special equipment to spread; good way to recycle.
good	good	rapid	Lightweight; can mix with other materials to improve appearance.
good	good	fairly rapid	Excellent winter cover mulch; cranberry vines have similar characteristics.
excellent	good	slow	Often used around strawberries and acid-soil plants, but can be used elsewhere.
fair; good where holes are cut for plants	good	moderate	In moist climates with long growing season, may need to add second layer; can bury under soil at end of season to speed decomposition.
poor, unless perforated	excellent	slow	Contributes nothing to the soil; must be handled twice a year; various colors available; some types specially formulated to be biodegradable.
good; doesn't mat	good	slow	Can be used year after year; weed- and pest-free; good for winter protection; now very hard to obtain.
fair	fair	slow, unless well weathered	Has high carbon content, so mix with nitrogen source to avoid depleting soil; slightly acidifies soil.
fair	good	slow	Provides nitrogen, potash, and micronutrients; excellent for sheet composting and as an insulating winter mulch.
good	fair, unless used over landscape fabric	extremely slow	Should be considered permanent mulch; slowly contributes some trace elements; use over landscape fabric where weeds are a problem.
good	good	fairly slow	Should be seed-free if possible; highly flammable when dry; spread thin, nitrogen-rich layer underneath to speed up decomposition and keep it from tying up nitrogen.
good	good	rapid	Needs to be replenished often; fairly acidic, so mix with lime.
good	good	slow	Won't tie up soil nitrogen if weathered outdoors for several months before use; if chips smell sour (like vinegar or ammonia), they must weather outdoors for a couple of months before they're safe to use on plants.
good	fair	rapid	Hardwood shavings better than pine or spruce; chips or sawdust make better mulch; use with nitrogen source to keep from tying up soil nitrogen; flammable.

Using Organic Mulches

Organic mulches can be used in all types of gardens, from vegetable beds to shrub borders and around fruit or ornamental trees. For ornamental beds that need periodic watering, consider installing a soaker hose. It can be covered with mulch and left in place all season. Snake the hose around the bed, leaving one end at the edge closest to the water source. Install a quick-coupling connector so it's easy to snap this end onto a regular hose.

HINT FOR SUCCESS

Fighting Slugs and Snails

Slugs and snails love organic mulches. If either is a problem in your garden, try waiting until late spring or early summer to mulch. You'll have to pull more weeds but you'll greatly reduce slug damage.

For vegetable gardens in all but hot climates, consider switching to black paper for weed control and use other methods to add organic matter. If you have small or raised beds, invest in copper barriers. Slugs and snails won't cross copper.

1 Calculate how much material you need. Remember that fresh wood chips and sawdust must sit outdoors for a few months to weather before use (water the piles if you don't get any rain). If you can't wait that long, spread a thin layer of nitrogen-rich material such as fresh grass clippings or blood meal underneath.

2 Prepare the garden before mulching. Remove any weeds and level or smooth the surface. If you're using a material that is slow to break down, spread a layer of finished compost or aged manure first to supply organic matter quickly.

How Much Do You Need?

Sometimes it's helpful to know the volume, rather than the weight, of material you need for mulching, topdressing, or amending your garden. One cubic yard equals 27 cubic feet. One bushel equals 1.25 cubic feet. A 5-gallon bucket filled to the top holds ⅔ cubic foot. *(For metric equivalents, see "Useful Conversions" on page 208.)*

Depth of Organic Material	Amount Needed to Cover 100 Square Feet
6 inches	2 cubic yards
4 inches	35 cubic feet (1¼ cubic yards)
3 inches	1 cubic yard
2 inches	18 cubic feet (⅔ cubic yard)
1 inch	9 cubic feet
½ inch	4½ cubic feet
¼ inch	2¼ cubic feet

Master Gardening Tip

Good Mulch Mixes

Mixing different types of mulches can give you a better cover. If a nearby restaurant can supply lots of used coffee grounds, mix 1 part grounds to 2 parts sawdust. That way, you can use fresh sawdust without temporarily robbing soil of nitrogen. Newspaper is very effective for smothering weeds. Improve its appearance by covering with bark chips or other attractive material that will also keep it from blowing in the wind. If you like the look of bark chips but can't afford much, spread chopped leaves or newspaper underneath and you'll only need a thin layer of chips.

3 Spread the mulch evenly over the surface. In general, spread it 2 inches deep around flowers and vegetables and 3 to 4 inches deep around shrubs and trees. (See mulch chart on pages 86–89 for specific recommendations.) Keep all mulch an inch or two away from plant stems to promote good air circulation and minimize diseases. If you've had problems with diseases, or if your climate is very humid, keep mulch even farther away.

4 Inspect your mulch periodically to make sure that wind or rain hasn't created a bare spot. Use a rake to redistribute it evenly. Eventually, mulches in ornamental beds will need replenishing. Be careful not to exceed the recommended depth when adding more.

In vegetable gardens, soft mulches can be tilled into the soil at the end of the season. Or you can remove the mulch before tilling and spread it over the soil afterward to protect against winter erosion.

Using Fabric, Paper, and Plastic Mulches

Landscape fabric, black paper, and plastic mulches are three varieties of inorganic mulch. All are great for suppressing weeds but should be used over a topdressing that supplies organic matter. They're all easy to install by following the steps below. Each has characteristics that make it particularly suited to specific uses.

Landscape Fabrics

Also called geotextiles, landscape fabrics are designed to be left in place for a long time. They let air and water through, so they're better around permanent plants than either paper or solid plastic. Several different types are available, ranging from plastic perforated with tiny holes to several layers of spun-bonded polyester. They must be covered with decorative mulch both because they're ugly and to make them last longer (direct sunlight causes them to disintegrate). Since they stay in place for several years, spread an inch or two of finished compost or aged manure before mulching. You won't be able to add organic matter once the material is in place.

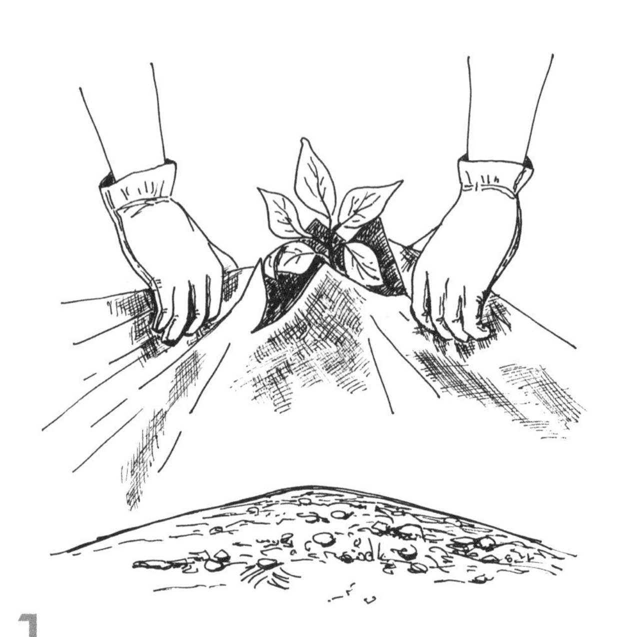

1 Measure out the area to be covered. Buy the widest rolls that will fit your site, as overlapping edges is somewhat difficult. Starting at one end of the garden, weight down the edge of the roll with rocks or bricks. Spread the material over the garden, working carefully around any existing plants. (It's easier — but not always possible — to plant after laying down this type of mulch.) For small plants, cut an X in the fabric and gently slide the X over the plants.

2 To lay material around large plants, cut a slit from the edge of the material to the plant, and cut a small circle or square large enough to fit around the stem(s). Spread the slit fabric around the plant, and continue spreading the fabric over the garden. If it takes more than one width to cover the area, overlap edges by 2 or 3 inches.

Plastic Mulches

Plastic (polyethylene) mulches are used to warm the soil and control weeds for row crops. Black is the most common color, but other colors are available. Red is particularly beneficial for tomatoes. Clear plastic warms soil quickly but doesn't control weeds. While plastic keeps soil from drying out, it lets in little or no rain, so plants may need drip irrigation. Plastic mulches aren't biodegradable (except for a few new and expensive types), so you have to remove them from the garden at the end of the season. Some are sturdy enough to be reused if they're stored in a dark spot over the winter.

Paper Mulches

Black paper mulch is a relatively new product designed for vegetable and annual gardens. Some people have used roofing (felt) paper, but the tar in it isn't good for people or gardens. Black paper costs about the same, contains no tar, and is biodegradable, so you don't have to remove it at the end of the growing season. (If you want it to completely decompose over the winter, cover it with soil — it's a bit hard to till under.) It controls weeds as well as black plastic and warms the soil almost as well (about 2–3°F, or 1–2°C, less).

3 Secure the edges of the material. In vegetable gardens, simply cover edges with soil; in windy areas you'll also need a few rocks or U-shaped wire pins. To plant, cut an X in the desired spot and fold back the flaps to expose soil. Try to keep soil off paper mulch and landscape fabric; spread an old fertilizer bag or piece of plastic when you're planting to catch loose soil.

4 Landscape fabric requires this additional step. Cover the entire area with an attractive mulch to hide the fabric from sight and from sunlight. Choose a relatively heavy, long-lasting mulch such as gravel or bark chips.

Soil Amendments

Soil amendments or conditioners are nothing more than material spread over and incorporated into the soil to improve structure, texture, or other properties. A few, such as manure and limestone, provide enough nutrients to be considered fertilizers as well. The overlap can be confusing, but few fertilizers improve the physical condition. Here we'll concentrate only on those amendments that improve tilth.

The use of soil amendments is ancient. Centuries ago, people learned that spreading manure over gardens (or letting animals onto fields to deposit their manure directly) improved plant health. Amending is an excellent soil-improving technique for new gardens. It also works particularly well in vegetable and annual beds that aren't in year-round cultivation. Around living plants, though, it's awkward at best.

When to Use Amendments

If most of the material has already decomposed, as with mature compost or well-aged manure, you can add it just days before planting. If you're using unfinished compost or material that hasn't yet broken down (newly fallen leaves or fresh manure), allow at least three weeks before planting. If possible, incorporate fresh material in fall so it has all winter to break down into usable form. This is especially important for carbon-rich materials such as sawdust.

DID YOU KNOW?

Amendment, Topdressing, or Mulch?

Anything that improves the condition of the soil can be called an amendment, but that's a bit confusing. Mulches and topdressings also improve the soil. It's easiest to use the word amendment (or soil conditioner) to refer to anything that's turned under the soil.

If something is left on top of the soil without being mixed in, it's either a topdressing or a mulch. If it's spread only once to supply nutrients or organic matter, it's a topdressing. If it's replenished to maintain a constant depth and used to hold in soil moisture or stop weeds, it's a mulch.

Leave the Lawn Clippings

Fresh grass clippings make a good soil amendment that breaks down quickly. But you can automatically topdress and fertilize your lawn if you recycle the clippings in place. Mulching lawn mowers chop grass finely so that it disappears quickly, contributing nitrogen, other nutrients, and organic matter to the lawn.

Leave the clippings where they're needed most. Collect them only once or twice a year (perhaps when you've let the lawn go a little too long) for use in compost or as a soil amendment. A 2- to 3-inch layer mixed into the soil will break down in a couple of weeks.

HINT FOR SUCCESS

Turning Under Garden Debris

Plan to turn under the remains of your vegetable or annual garden at the same time you incorporate amendments. You'll save frustration and use less effort in the long run if you chop long stems or vines first. Make a few passes with a string trimmer for larger areas, slicing off a few inches with each pass. If you've got only a few plants, you may find it easier to cut them into sections with hand pruners. Chop the material before spreading your amendment. Then you can turn both under at once.

If you won't be planting early crops in an area, you can leave both plant residues and amendments on top of the soil as a winter mulch. Turn both under in spring, waiting until the soil dries out enough to work. The soil should be settled sufficiently by the time you set out frost-tender plants such as tomatoes and eggplants.

Using Soil Amendments

Most amendments work best if you incorporate them into soil in fall to prepare for planting the following spring. If the materials are well decomposed, you can mix them in shortly before planting.

Sawdust, wood shavings, and other carbon-rich materials require added nitrogen to reach the optimum carbon-nitrogen balance for decomposition. If you have time, leave them out to weather before spreading on soil. If you want to use them fresh, mix them with a smaller amount of a good nitrogen source, such as grass clippings. (See also page 56.)

Microbes will scavenge much of the soil's existing nitrogen as they break down carbon (which is luscious food to them). Eventually, this nitrogen will be returned to the soil, but that's not much help to this year's plants. When you add nitrogen before spreading sawdust, you ensure there's enough nitrogen left over for this year's plants.

HINT FOR SUCCESS

Use a garden fork to mix small areas. If the material is relatively fine, a small mechanical cultivator can handle it. For coarse, unfinished compost and unchopped hay or straw, use a larger rototiller.

1 Calculate how much material you will need (see page 96). Unload the material in several equal-size piles spaced evenly over the entire bed. Rake out each pile to an even thickness, taking care to cover any gaps.

2 Mix the material lightly into the top 4 to 6 inches of soil, just enough to keep it from forming a separate layer that could mat down. The idea is to put the amendment within reach of earthworms, fungi, and bacteria.

3 Don't rake the soil smooth until just before planting. Uneven soil with large clods resists erosion better. If you're working in fall, cover the soil with at least a thin layer of organic mulch to protect it until spring.

Types of Soil Amendments

All of the softer or finer mulch materials also make good soil amendments. Hard materials take too long to break down. Any materials that contribute organic matter provide the following benefits: improved drainage, loosened clay and compacted soils, improved nutrient and water retention in sandy soils, and supplies of micronutrients as well as traces of major nutrients.

A few materials that contribute organic matter aren't listed here because they're too concentrated. Poultry manure and bat guano should be treated as fertilizers rather than soil amendments, to prevent problems with excess nutrients. The last two materials listed here contribute no organic matter but condition the soil in other ways.

The amounts recommended here are for new gardens and soils low in organic matter. If your organic matter levels are about average, use half as much. Once your soil is really rich and fertile, you can use even less as a maintenance diet. In warm, humid climates, use a bit more than half this amount as a maintenance diet.

If your material is coarse and loose (such as unfinished compost), aim for the deeper end of the thickness range. If it's very fine (such as dehydrated manure), spread it more thinly. Depth isn't as critical for amendments as it is for mulches. *(For metric equivalents, see "Useful Conversions" on page 208.)*

Material	Amount for 100 Square Feet	Benefits
Compost, homemade or commercial	100 lbs. (moist), 15 lbs. (dry); 2–4 in. thick	Excellent source of organic matter; finished compost can be used anytime; unfinished compost should be turned under in fall or several weeks before you wish to plant.
Manure	50–100 lbs. (moist), 15 lbs. (dry); about 2–4 in. thick	Excellent source of organic matter; supplies many nutrients so reduce fertilizer applications; well-aged or composted manures can be added anytime, but fresh and dried manures should be turned under in fall or several weeks before planting; cow and steer manure can contribute to salt build-up, so they're not good for salty soils.
Leaf mold	50–100 lbs.; 2–4 in. thick	Excellent source of organic matter; can be used anytime.
Chopped leaves	15 lbs.; 4 in. thick	Excellent source of organic matter; turn under in fall or several weeks before planting; good (low-level) source of many nutrients.
Fresh sawdust	15 lbs.; 2–3 in. thick	Good source of organic matter; very high in carbon, so compost first or add nitrogen (blood meal or fresh grass clippings) at same time; turn under in fall or a couple of months before planting.
Hay or straw (as weed-free as possible)	15 lbs.; 2–3 in. thick if chopped	Good source of organic matter; easier to turn under if chopped or after it's been used as mulch; turn under in fall or several weeks before planting; high in carbon, so add a little nitrogen to help it break down quickly.
Peat moss	4 lbs.; 1 in. thick	Supplies organic matter; acidifies soil, so use only for acid-loving plants or to help lower soil pH; mix well into soil, as peat moss repels water if left on soil surface; can soak up lots of moisture, so water soil after turning under.
Apple or grape pomace	100 lbs.; 2–3 in. thick	Good (low-level) source of potassium and phosphorus; often free from local producers; may contain pesticide residues; very moist unless composted first, so heavy and somewhat awkward to handle.
Greensand	2 lbs. a year or 4–5 lbs. every 2–3 years	Supplies no organic matter; helps loosen clay soils; improves water and nutrient retention in sandy soils; rich, slow-release source of potassium and micronutrients, so don't add potassium for 3 years after applying; good source of growth-promoting silica.
Gypsum	2 lbs.	Supplies no organic matter; corrects soil structure problems caused by too much magnesium or sodium; may help loosen clay soil; supplies lots of calcium (without changing pH), so don't use where calcium levels are already high; not good for very acidic soils (those with pH below 5.8).

Using Topdressings

The simplest way to add organic matter is to spread it on top of the soil, which is called topdressing. Any of the materials used as amendments will do. Topdressing works slowly — you have to depend on the soil organisms to do the mixing and get the material into the root zone. You can speed up the process by using only well-decomposed materials such as fine compost and worm compost. Given time, earthworms and other organisms will do an excellent job of incorporating the material, saving you a lot of effort (yet another reason for maintaining an active community of organisms in your soil!).

MASTER GARDENING TIP

Where to Use Topdressings

This method is particularly suited to established beds of perennial flowers and perennial crops such as asparagus and rhubarb. It also works well for plantings of blueberry bushes and ornamental shrubs whose shallow roots could be damaged by digging. If the plants are deep-rooted, you can scratch the surface soil to help mix in the topdressing, which will speed its use by plants.

Topdressing is great for giving annual crops and flowers an extra boost once they've started growing. Spread a well-decomposed form of organic matter in strips along rows, or in a circle around individual plants. (If you're using mulch, pull it out of the way first.) This method of topdressing is usually referred to as sidedressing.

Green Manures: Homegrown Organic Matter

A green manure is a crop grown for the primary purpose of turning it under to supply the soil with organic matter. Some green manures offer additional benefits, such as weed control and disease reduction. Many farmers rely on them because they're cost-effective on a large scale, but backyard gardeners often don't realize how easy they are. You grow your own source of organic matter right where you need it.

Green manures improve soils in several ways. Their roots can reach much farther than any rototiller or plow — and also far enough into the subsoil to scavenge for nutrients. When the plants decompose, these nutrients are returned to the topsoil layer, where even shallow-rooted plants can reach them. As the roots decay, they leave organic matter and fine channels deep in the soil.

Save Money and Effort

If you don't like shoveling amendments or making compost, using green manures may be your method of choice, at least for vegetable and annual beds. They're cheap because you buy only the seeds. Don't look for them in fancy packets at the garden center; buy them by the pound and you'll save money. You don't need any special equipment, either. While a string trimmer is handy for cutting plants into smaller pieces and a rototiller is handy for turning under large areas, you can easily turn under smaller areas with a garden fork. All you need is to set aside some time for growing green manures, either before, after, or in between your other crops.

DID YOU KNOW?

Why Grow Green Manures?

- Enhances biological activity in soil.
- Adds *lots* of organic matter to soil.
- Increases availability of existing nutrients.
- Keeps nutrients from washing away.
- Reduces need for nitrogen fertilizers (legume types).
- Improves soil tilth.
- Reduces weeds.
- Reduces pests and diseases when used in crop rotations.
- Inexpensive.

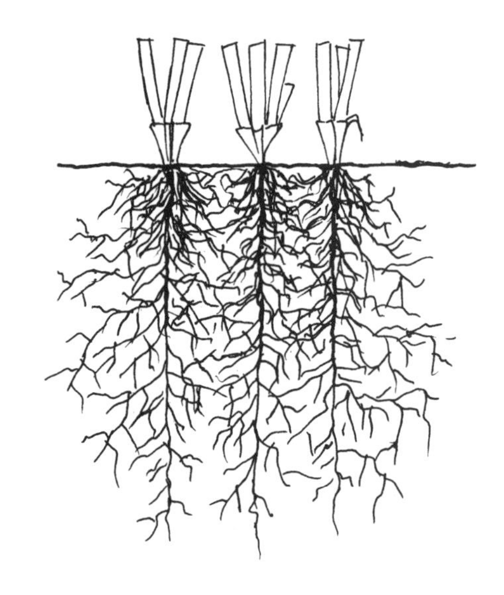

Alfalfa roots can easily reach 9 feet deep in good soils, though most reach 5 feet down. Red clover and soybean roots can grow 5 or 6 feet deep, though most go down about 3 feet. Even grains and grasses can grow long roots in good soils, reaching from 2½ to 5 feet down. The abundant, fine roots of grains and grasses make them good overall structure improvers.

Legumes and Grasses

Two general types of plants are grown as green manures: legumes and grasses. Legumes are members of the plant family that includes beans, peas, and all of their relatives, such as clover and alfalfa. Grasses (in a family of their own) include all of the cereal grains; oats and rye are most commonly used for green manures. All common green manures belong to these two families except buckwheat (a relative of rhubarb) and rapeseed (in the cabbage family).

DID YOU KNOW?

Green Manure or Cover Crop?

Some people use the terms cover crop and green manure interchangeably. When a green manure crop is grown for the added benefit of reducing soil erosion, it's called a cover crop. Choose whichever name fits your primary purpose.

Legumes Supply Nitrogen

Legumes are particularly valuable because they release nitrogen after they're turned into the soil. Legumes can easily supply half of the nitrogen requirements of plants grown the following year. With very careful management (which includes fertile soil, good growing conditions, and use of the right inoculant), they can supply all nitrogen needs.

A special kind of bacterium forms colonies, or nodules, on the roots of legumes. These bacteria work in partnership with the plants' roots. Together, they transform otherwise unusable nitrogen from the atmosphere into ammonia, a form plants can use readily. Scientists call this process nitrogen fixation and the organisms responsible nitrogen-fixing bacteria. To ensure that enough bacteria are present to fix abundant nitrogen, legume seeds (or the soil) are treated before planting with inoculant (a culture of the type of bacterium that forms on that specific legume).

Turning under legumes at any growth stage provides organic matter and enhances soil life. To get the full benefit of nitrogen fixation, though, allow legumes to grow for at least a full season. Perennial plants such as alfalfa, birdsfoot trefoil, and red clover produce even more nitrogen if allowed to grow for two years. During those two years, they can be mowed several times. The mowings make great mulch or compost.

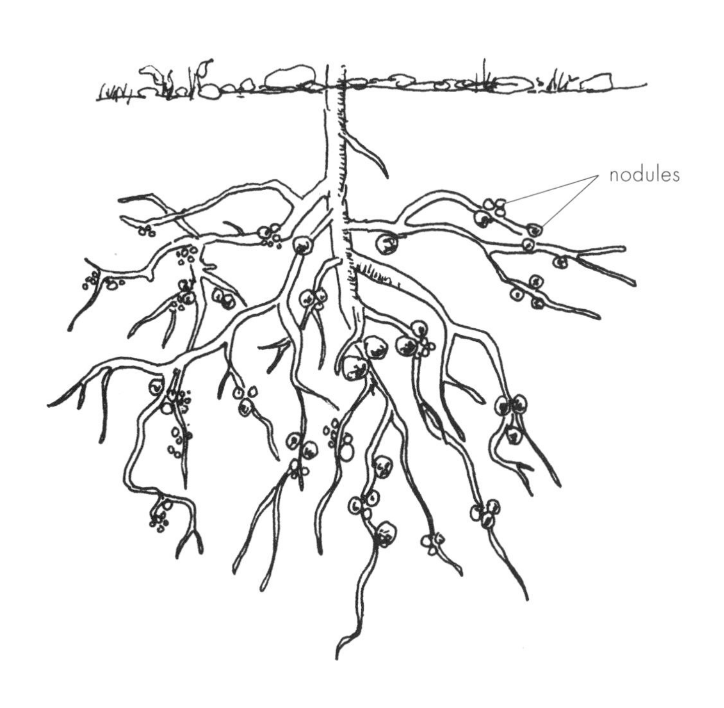

Uproot a healthy pea or bean plant and look at its roots. The round nodules are small, often the size of a large pinhead, and white to pinkish in color. They should be scattered throughout the entire root system. The more nodules you see, the better. More nodules means more nitrogen fixation (and better plant and soil health).

Common Legumes for Green Manures

The following chart will help you choose the best legume green manure for your situation. While these seeds are usually sold by common name, botanical names are also included because different common names are sometimes used for the same plant. The second column describes which soil conditions are best for each plant. Amounts of seed to use are given for areas of two different sizes; if your garden is even larger or smaller, increase or decrease the amounts correspondingly.

Type	Soil	Oz. of Seed/ 100 Sq. Ft.	Lbs. of Seed/ 1,000 Sq. Ft.	Planting Depth (In.)
Alfalfa *(Medicago sativa)*	Well-drained and neutral to alkaline (pH 6.5 or more); needs lots of P and K	¾	½	½
Beans, fava (broad) *(Vicia faba)*	Not too acidic (pH at least 6)	6	4	2
Beans, mung *(Vigna radiata)*	Most soils if not too acidic (pH at least 6)	4½	3	1
Beans, soy *(Glycine max)*	Prefers loam and neutral pH; tolerates poor drainage	6	4	1½
Birdsfoot trefoil *(Lotus corniculatus)*	Tolerates poor or shallow soils, poor drainage	¼–¾	¼–½	½
Clover, alsike *(Trifolium hybridum)*	Tolerates heavy, wet, and/or acidic soils	¾	½	½
Clover, crimson *(Trifolium incarnatum)*	Neutral pH, well-drained; tolerates poor soil	¾–1½	½–1	½
Clover, red *(Trifolium pratense)*	Average; prefers pH of 6 but tolerates slightly lower	¾	½	½
Clover, white Dutch *(Trifolium repens)*	Average; tolerates some drought	¼–½	¼	½
Cowpeas *(Vigna unguiculata)*	Well-drained; tolerates low fertility	4½–6	3–4	1½
Field peas *(Pisum sativum)*	Tolerates heavy soils	4½–6	3–4	1½
Lespedeza (bush clover) (*Lespedeza* spp.)	Tolerates poor and/or acidic soils	1½	1	½
Lupines (*Lupinus albus* and other spp.)	Tolerates acidic soils	3–4½	2–3	1
Sweet clover 'Hubam' *(Melilotus alba)*	Well-drained and neutral; tolerates less fertility than alfalfa	¾	½	½
Sweet clover, white or yellow (*Melilotus* spp.)	Well-drained and neutral; tolerates less fertility than alfalfa	¾	½	½
Vetch, hairy *(Vicia villosa)*	Most soils; tolerates poor and/or acidic soils	2½	1½	¾
Vetch, common *(Vicia sativa)* and Hungarian *(V. pannonica)*	Needs fairly fertile soil; tolerates some acidity	2½	1½	¾

Follow the recommended planting depth to ensure good germination. Check the "Best Regions" column to make sure a specific type is suited to your area. The last column includes tips for specific uses and best planting times. *(For metric equivalents, see "Useful Conversions" on page 208.)*

Best Regions	Comments
All	Perennial; deep roots help loosen soil; best nitrogen fixer but must grow a full year (mow 2–4 times/year for hay or compost); plant in spring or late summer; turn under the following spring or fall.
All	Annual; sow in spring or summer; turn under in summer or fall; germinates easily; tolerates poor growing conditions; beans are edible. (Any edible beans can be turned under after harvest.)
South	Annual; needs warm weather and warm soil; sow in late spring or summer; turn under in summer or fall.
All	Annual; needs warm weather; sow in spring or summer; turn under in summer or fall; good summer crop in Deep South.
All	Perennial; good where soils are too poor for alfalfa; grows slowly at first; sow in spring; turn under in fall or the following year.
All except southern and southeastern states	Annual; tolerates poor drainage and acidic soils better than other clovers; sow in spring to late summer; turn over in fall or spring.
All	Annual; where winters stay above 10°F (-12°C), sow in fall and turn under in spring; in North sow in spring and turn over in fall; very attractive.
Northern and central states	Short-lived perennial; good phosphorus accumulator; sow from early spring to late summer and turn under in fall or the next spring; fast-growing; 'Mammoth' is a good variety.
All	Perennial; tolerates foot traffic; shorter than most green manures, so good for paths or living mulch under existing crops; sow from spring to summer; turn under in fall or spring.
Southern and central states	Annual; needs warm soil; drought tolerant; also tolerates some shade; sow in late spring to early summer; turn under in summer or fall.
All	Annual; works best if combined with oats or other grain; sow in spring and turn under before peas fill out pods; where winters are mild, sow in fall and turn under in spring.
Southeast	Annual (except for perennial *sericea*); good for restoring eroded soils; three varieties are all treated the same way, but one *(sericea)* is also good for Southwest; sow in early spring and turn over in summer or fall.
All	Annuals; white, blue, and yellow varieties all treated the same way; sow in spring and turn under in summer (before pods form); in mild climate can sow in late summer and turn under in spring.
Northern and central states	Annual; almost as good as alfalfa for producing nitrogen; plants grow 6 feet (1.8 m) tall or more, producing loads of organic matter; sow in early spring and till under in fall.
All but Gulf states	Biennial; deep roots loosen soil; almost as good as alfalfa for producing nitrogen; plants grow as tall as 'Hubam'; sow in early spring and till under in fall; white type is hardy, so can also be sown in late summer; yellow type best for Southwest; excellent plant for honeybees.
All	Biennial; very hardy; sow in spring or fall and till under the following fall or spring; for winter cover crop, mix with winter rye, oats, or winter wheat; best choice for short-term nitrogen fixation.
All	Biennial; sow in spring or fall and till under the following fall or spring; not as hardy as hairy vetch so don't sow in fall where winters are severe; Hungarian not as good for Gulf states.

Grasses and Other Nonlegumes for Green Manures

The following chart will help you choose the best nonlegume green manure for your situation. While these seeds are usually sold by common name, botanical names are also included because different common names are sometimes used for the same plant. The second column describes which soil conditions are best for each plant. Amounts of seed to use are given for areas of two different sizes; if your garden is even larger or smaller, increase or decrease the amounts correspondingly.

Type	Soil	Oz. of Seed/ 100 Sq. Ft.	Lbs. of Seed/ 1,000 Sq. Ft.	Planting Depth (In.)
Barley *(Hordeum vulgare)*	Neutral to alkaline; not good for sandy soils	4	2½	¾
Bromegrass *(Bromus inermis)*	Average; widely adaptable	1½	1	½
Buckwheat *(Fagopyrum esculentum)*	Tolerates poor and/or acidic soils	3–4½	2–3	¾
Millet *(Panicum miliaceum)*	Prefers good drainage; tolerates acidic soils and drought	1½	1	¾
Oats *(Avena sativa)*	Prefers cool soil and even moisture; tolerates acidic and alkaline soils; avoid heavy clays	4½–6	3–4	1
Rapeseed (canola) *(Brassica napus)*	Average	7½	5	½
Rye, winter *(Secale cereale)*	Widely adaptable; tolerates drought	4½–6	3–4	¾
Ryegrass, annual (Italian) *(Lolium multiflorum)*	Prefers good drainage, but fairly adaptable	1½–3	1–2	¾
Sudan-grass (grain sorghum) *(Sorghum sudanense)*	Tolerates alkaline or poorly drained soils; also tolerates drought	1½–3	1–2	¾
Wheat *(Triticum aestivum)*	Fertile; neutral to slightly alkaline	4½–6	3–4	¾

smooth bromegrass

Follow the recommended planting depth to ensure good germination. Check the "Regions" column to make sure a specific type is suited to your area. The last column includes tips for specific uses and best planting times. *(For metric equivalents, see "Useful Conversions" on page 208.)*

Best Regions	Comments
All but Gulf states	Annual; use spring varieties in the North, winter varieties in milder climates; sow in spring and turn under in summer, or sow in fall and turn under the following spring.
Northern and central states	Perennial; easy to grow; very cold tolerant; hardier and more heat tolerant than rye.
All	Annual; fast-growing; can be turned under in 5–6 weeks; not frost hardy and needs warm soil to grow well; plant anytime after last spring frost and after soil is warm until midsummer; turn under when first flowers show; excellent for smothering weeds; accumulates phosphorus.
Northern and central states	Annual; fast-growing; good for smothering weeds; needs warm weather; sow in late spring or summer, turn under in summer or fall.
All	Annual; fast-growing; good mixed with slow-growing legumes for quick cover; sow in spring through summer and turn under before seeds form (6–8 weeks), or sow in late summer and turn under in spring; not cold hardy, but frost-killed plants can still provide winter cover.
Northern and central states	Annual; fast-growing; sow from spring through summer, turn under before pods form; good honeybee plant if allowed to bloom (but don't let it form seeds, unless you want to press your own canola oil and are willing to pull up seedlings); greens are edible; in the South, use winter rape and plant in fall for spring tilling.
All but Deep South	Annual; hardiest winter cover crop; can be sown later than others; plant in fall and turn under in spring; turn under 3–4 weeks before sowing other seeds (don't need to wait with transplants); in cooler climates, can also sow in spring and turn under in fall.
All	Annual; fast-growing; sow from spring through summer and turn under before plants bloom or form seeds, or sow in early fall and turn under in spring; not cold hardy but frost-killed plants can still provide winter cover.
All	Annual; very tall (like corn); thrives and grows rapidly in heat; good for smothering weeds and controlling erosion; sow as soon as soil is warm in spring through summer, turn under when plants reach 6 inches.
All	Annual; sow winter varieties in late summer to fall and turn under in spring (not as hardy as winter rye); sow spring types as early as possible in spring and turn over before they bloom.

annual ryegrass

Growing Green Manures and Cover Crops

There are two secrets to success with green manures:

- **Choose the right plant for your site, season, and primary purpose** (building nitrogen, suppressing weeds, winter cover). Most legumes require neutral soil and the same phosphorus and potassium levels needed to grow good vegetables; you may have to spread lime or fertilizer before you plant.
- **Turn under the crop at the right time.** For winter cover crops, the right time is as soon as the soil is dry enough to work in spring, or at least two weeks before planting. Turn under other crops before they form seeds.

If you have a small garden, remember that a pound of seeds goes a long way. With smaller seeds, you can buy a pound and use it for a couple of years. Larger seeds such as peas and beans don't keep as well, so buy fresh seeds every year. Mixing old seeds of garden bean and pea varieties with your green manure seeds is a great way to use them up.

HINT FOR SUCCESS

Legume Inoculants

The bacteria responsible for fixing nitrogen are fussy. Each type of inoculant works for only a few species of legumes.

You can usually buy a mix designed for any peas, snap or dry beans, and lima beans. You need a different type for fava (broad) beans and vetches. Soybeans require their own inoculant.

Another species works for all true clovers (ladino, alsike, white Dutch, crimson, red, 'Mammoth'), but for sweet clovers you need a different kind. When in doubt, ask your supplier.

1 Calculate how much seed you'll need. Order seed ahead of time if it's not available at your local garden center or farm store. If you're planting legumes, also purchase the right inoculant (see box above).

2 Dig up and remove any sod and weeds. Loosen the top few inches of soil. If the soil is poor, amend with organic matter and fertilizer (or choose a green manure that tolerates poor soil). This is a great time to apply a slow-release nutrient source such as granite meal, greensand, rock phosphate, or limestone. Rake the seedbed smooth.

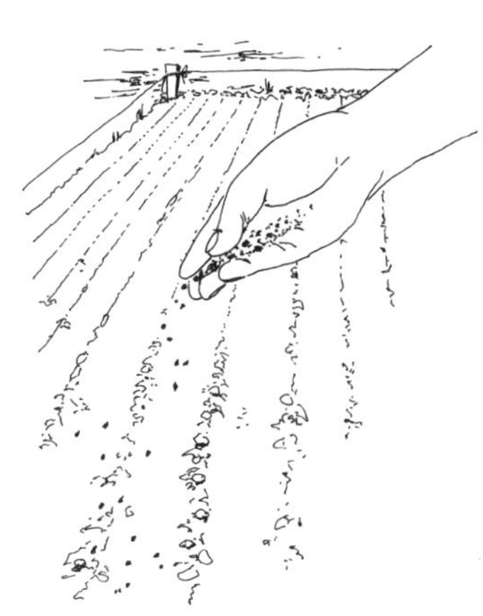

3 Scatter seeds evenly over the entire bed. For large areas, a broadcast seeder is a big help. If you're sowing less than a pound over 1,000 square feet, mix seeds with sand or sifted soil to get even coverage. Rake the bed to cover seeds and tamp down the soil. Spread a light covering of straw or grass clippings to reduce the need to water. (Floating row covers also reduce watering needs.)

4 If lots of weeds sprout among slower-growing perennial legumes, mow or cut your planting once it reaches a few inches high. The perennials will bounce back quickly and overtake the annual weeds. If you're leaving a crop in place for a whole year, mow periodically to keep the plants soft for easy cutting. Use the mowings as mulch or in a compost pile.

5 Cut or mow the crop before tilling. A string trimmer makes quick work of soft, lush growth. For tall crops, make several passes with the trimmer, to chop while you cut. A sickle-bar mower or scythe will cut tougher stems. Before turning under, leave the cut material for a couple of days to wilt.

6 Turn under the cut material and mix into the top few inches of soil. For large areas, use a rototiller (small cultivators work only if material is chopped finely). Vigorous perennials such as winter rye may need to be turned again in a week to stop resprouting. After turning under, leave soil surface rough. The bed should be ready for smoothing and planting in a week or two. Wait three to four weeks if you turned under a large amount of coarse material (full-grown alfalfa) or if you'll be planting seeds after turning under winter rye (which temporarily suppresses germination).

Green Manures for Special Purposes

All green manures are good for adding organic matter. By choosing plants carefully, you can get increased or specialized benefits. Combining different plants creates multipurpose green manures. The suggestions here have been worked out over many years and in many different regions.

Suppressing Weeds

Two plants are outstanding for weed control: buckwheat and rye. Buckwheat works by growing rapidly, expanding its large leaves to form enough shade to stop common annual weeds from sprouting. Growing two crops in a single season is a great way to prepare new beds or new lawns. You can't let the buckwheat go to seed, though, or you'll be pulling it instead of other weeds.

Winter rye is a natural herbicide. It produces chemicals that are toxic to many weed seedlings. It works for most perennial as well as annual weeds. One crop is enough for significant weed reduction. Wait three to four weeks after turning under to sow seeds of other crops to ensure good germination.

The best legume for smothering weeds is field peas. While the results aren't as dramatic as with buckwheat or rye, you get the added benefit of nitrogen fixation.

For really weedy areas, or to control pesky perennial weeds such as couchgrass, plant both buckwheat and winter rye. As soon as frost danger has passed, plant buckwheat thickly over the area. When the buckwheat starts to bloom, mow it and turn it under. Immediately plant a second crop of buckwheat. When this crop blooms, mow it and turn it under. Immediately plant winter rye, and leave it to grow over the winter. As soon as the soil can be worked in spring, turn under the rye. You'll be amazed at how rich and weed-free your soil is.

MASTER GARDENING TIP

Mixing for Maximum Benefit

Grasses provide fast cover and legumes provide nitrogen, so mixing them gives you both benefits. Dr. Matt Leibman, of the University of Maine, has designed an easy-to-grow mix that provides both quick cover and nitrogen fixation. Containing 60 percent field peas, 20 percent oats, and 15 percent hairy vetch, it's used at a rate of 5 pounds per 1,000 square feet (½ pound for 100 square feet). Sow in early spring and turn over in summer or fall, or sow in late summer and turn under the following spring.

For a winter cover crop, a mix of hairy vetch and winter rye is hard to beat. Both are very hardy and can be sown until mid-fall even in cold northern areas.

If you're mixing two or more types of plants, reduce the seeding rate accordingly (but stay a little on the generous side). For two plants, use just over half the recommended amount of each.

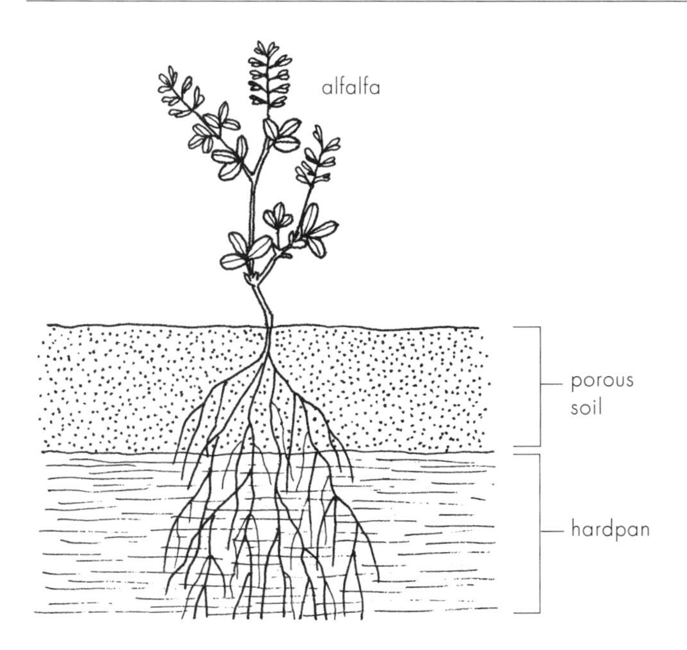

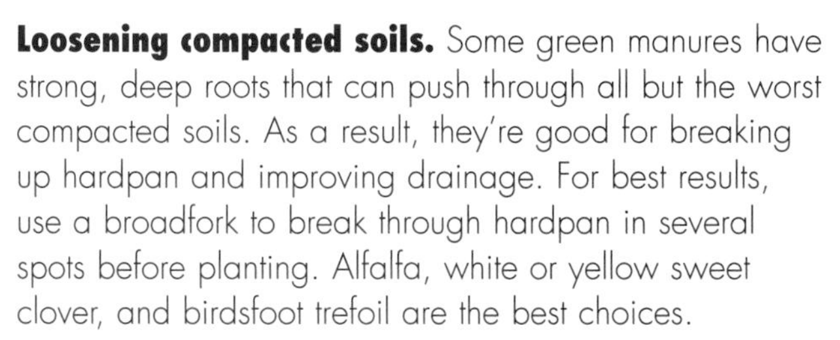

Loosening compacted soils. Some green manures have strong, deep roots that can push through all but the worst compacted soils. As a result, they're good for breaking up hardpan and improving drainage. For best results, use a broadfork to break through hardpan in several spots before planting. Alfalfa, white or yellow sweet clover, and birdsfoot trefoil are the best choices.

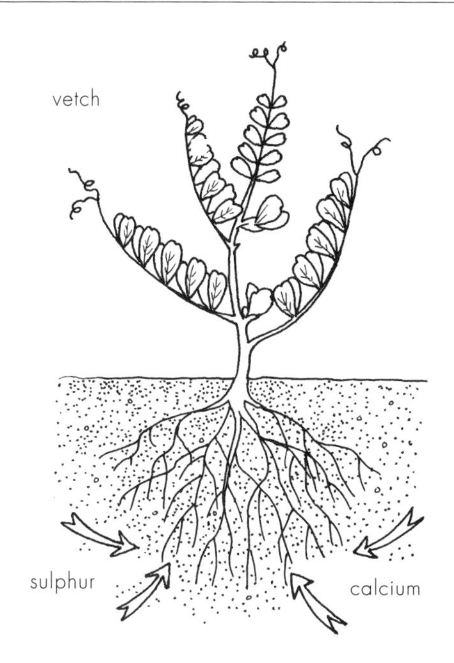

Concentrating nutrients. A few plants not only bring up nutrients from the subsoil, they also concentrate them. Growing them gives you measurable (though not huge) increases in the topsoil's available nutrient levels. They're called accumulators. Buckwheat increases calcium and phosphorus levels. Legumes, especially red clover, are good phosphorus accumulators. Vetches can increase levels of both sulfur and calcium.

Living mulches. If you don't want to take beds out of circulation for a season to grow green manures, try growing them at the same time as your other crops. Oats can be sown underneath existing crops, or along with tall crops, from mid- to late summer. (Don't plant oats earlier or they'll go to seed.) Leave in place until the following spring to provide good winter cover. White Dutch clover only grows a foot (30 cm) tall, so it's the best legume for growing with larger vegetables. (Corn is so tall that soybeans can be grown as a companion.) You can plant clover between rows of young squash or right around the bases of corn and members of the cabbage family, once they reach 6 to 8 inches (15 to 20 cm) tall. It also works well around fruit trees and to cover paths between garden beds.

Winter rye and hairy vetch need to be planted by mid-fall in order to develop enough growth before cold weather. Sow them right around late crops such as broccoli.

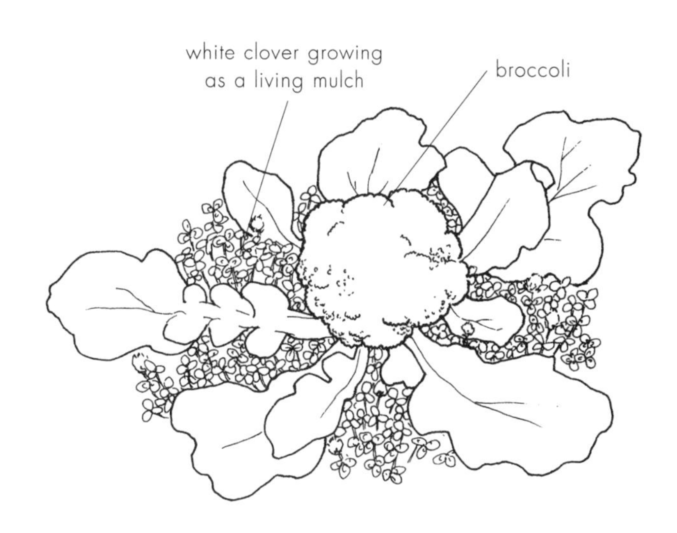

Combination Cover Crop and Mulch

U.S. Department of Agriculture researchers in Beltsville, Maryland, have produced amazing results growing tomatoes with no nitrogen fertilizer, using just hairy vetch as a combination cover crop and mulch. Yields are 100 percent higher than in fertilized, bare soil! Plus, there are few, if any, weeds. Researchers are trying this with other crops.

To get these benefits, plant hairy vetch in fall about a month before the first hard frost (usually one to three weeks before the first fall frost). Use 1 ounce of seed for each 10 square feet of bed. The first year, you'll need to use an inoculant that's specific for vetches. The vetch should reach about 4 inches (10 cm) tall before it stops for the winter. In spring, it'll start growing again, reaching 3 to 4 feet (90 cm to 1.2 m) and beginning to bloom by the time it's warm enough to set out tomato plants.

The day before setting out your tomato plants, cut the vetch back to 1 or 2 inches (2.5 to 5 cm). You'll need hedge shears, a hand sickle, or a string trimmer. Lay the cut vetch on the ground to make a soft, dense mulch 4 or 5 inches (10 to 13 cm) thick. Push aside the mulch to dig planting holes. The vetch must stay moist in order to decompose and release nitrogen. That means if it's dry enough to water your tomatoes, make sure to water the mulch as well (set soaker hoses on top of, not underneath, the mulch).

By the end of the season, most of the mulch will have disappeared. Cut, remove, and compost the tomato plants. Don't till the soil — one theory for the success of this system is that the lack of tilling results in improved soil structure and more numerous earthworms and other organisms. Just scratch the surface with a rake before planting the next batch of vetch.

HINT FOR SUCCESS

Ready, Set, Sow!

If your soil takes a long time to dry out in spring or you want to grow extra-early crops, try oats for a winter cover crop. Since oats are killed back in winter, all you have to do is rake or push them aside in spring and you're ready to sow. This allows you to get peas into the ground even before the soil is ready to till.

Sow oats in late summer to early fall, at least a month before hard frost, so they'll grow enough to provide good cover. Make a level seedbed for the oats so you'll have a nice seedbed for your spring crop. In spring, rake the partially decomposed oats off the bed. Plant as soon as the soil is warm enough for the crop you're growing.

Grow exceptional tomatoes and grow your own mulch at the same time by using hairy vetch as a cover crop. As a bonus, you won't have to weed as often and won't need as much fertilizer. You'll have to plan ahead if you combine this technique with crop rotation, though; in fall, sow hairy vetch in the area where you'll grow tomatoes the following year.

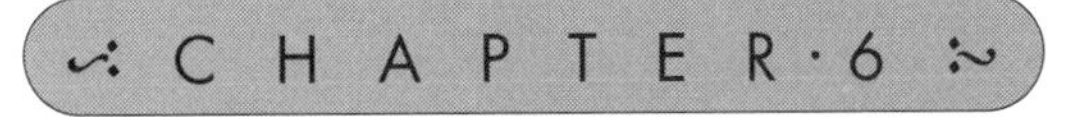

How to Choose and Apply Fertilizers

Many people think fertilizers are the basis of good fertility, but they're really just the tip of the iceberg. Pouring on fertilizer won't guarantee rich, fertile soil. Fertilizers won't make up for poor drainage or compacted soil. Fix the drainage or structural problems and you'll find plants suddenly make better use of any fertilizer. The physical factors mentioned in chapter 1 that promote good tilth also promote the best, most efficient fertilizer use.

Test Before You Fertilize

Creating a good balance of readily available nutrients is the next step in building rich, fertile soil. Start by testing your soil, as described on pages 22–23, to see what nutrients it needs and whether there are pH problems that interfere with nutrient use. Correct acidic or alkaline soil (see chapter 7) before adding fertilizer.

Feeding and Maintaining

Once you understand what your soil needs, this chapter will help you find the best sources of those nutrients. There are fertilizers to correct single or multiple deficiencies. Some can also help make soil more acidic or alkaline. You save money by buying only those nutrients your soil needs, instead of relying on standard mixtures for every situation.

Once you've achieved well-balanced, fertile soil you can switch to a reduced, maintenance diet of a general-purpose fertilizer. You'll still have to add organic matter each year. You'll also need to test soil periodically and amend it as needed to keep the pH around 6.5.

In This Chapter

Organic vs. Synthetic Fertilizers

Fertilizers either come from organic sources or are synthesized. Some say that all synthetic fertilizers are bad, and others say there's absolutely no difference, but it's not that simple. All types of fertilizers can help your garden; some just require more careful use than others.

It's not really the source of the fertilizer that determines whether it must be used with care. It's whether the fertilizer is both concentrated and fast-acting. These qualities apply to most popular synthetic fertilizers but also to some that are classified as organic (such as nitrate of soda).

Concentrated and Fast-Acting Fertilizers

Concentrated fertilizers have both good and bad characteristics. Concentrated means you don't need to haul as much fertilizer over to your garden to supply a given amount of nutrient. But the same high concentration means it's easier to overdose. Too much concentrated fertilizer can upset the soil balance so much that vegetables yield less than they would have with no fertilizer! And a small excess of some micronutrients is enough to kill sensitive plants.

If fertilizers are fast-acting, it means they're easily dissolved in soil. That's also both good and bad for the gardener. Plants can use the nutrients right away, but dissolved nutrients may be washed out of soil by heavy rains. If you want these nutrients to stick around in soil for any length of time, you have to maintain ample levels of organic matter.

When fertilizers are both concentrated and fast-acting, there's an additional danger. If they come in direct contact with a plant, they can burn its leaves, stems, or roots. They can injure soil organisms, too. When soil organisms are surrounded by a sudden, intense burst of one or more nutrients, they stop processing organic matter, which stops the release of other nutrients.

Organic Fertilizers

Fertilizers that are only moderately concentrated and release nutrients more slowly over a longer time give gardeners more leeway to make mistakes without seriously harming either soil or plants. That's one of the main benefits of using organic fertilizers, as most have both of these characteristics. Fertilizers from organic sources have other benefits, too. Many supply several important micronutrients missing from traditional synthetic fertilizers, and some supply organic matter.

DID YOU KNOW?

Advantages of Organic Fertilizers

- Nutrients are usually released more slowly.
- They last longer in soil, supplying nutrients over a longer time.
- There's less danger of burning plant roots than there is with concentrated synthetic types.
- There's less danger of oversupplying one nutrient, upsetting the soil balance and causing deficiencies of other nutrients.
- There is less danger of toxic salt buildup.
- These fertilizers enhance the activity of soil organisms.
- They are less apt to burn or shock soil organisms.
- They supply micronutrients.
- They usually require less energy (fuel) to produce.
- Often, they're less expensive.
- There's less danger of nutrients washing away to pollute ground- or surface water.

A Dictionary of Terms

Fertilizer labels can be confusing. "Balanced" doesn't mean that nutrient levels are balanced; it just means a fertilizer contains the big three: nitrogen (N), phosphorus (P), and potassium (K). Seeing "organic" on the label won't guarantee that all ingredients are derived from natural animal, plant, or mineral sources.

Here's a quick guide to different types of fertilizers. Most fall into more than one of the following categories.

- **Simple fertilizer** contains one major nutrient (N, P, or K).

- **Compound fertilizer** provides two or more of the major nutrients.

- **"Complete" or balanced fertilizer** contains N, P, and K (5-10-5 or 5-10-10); may not contain any micronutrients; may be either synthetic or organic.

- **Special-purpose fertilizer** is usually a balanced fertilizer, but one that is formulated for particular types of plants rather than for general use. (Bulb fertilizers and formulas for acid-loving plants are examples.)

- **Slow-release fertilizer,** also called **timed-release fertilizer,** is a pelletized form of concentrated, fast-acting fertilizer treated to make it release nutrients (especially nitrogen) gradually and over a longer period of time. Such formulas won't burn plants or soil organisms. More expensive than ordinary types. Also available as spikes, tablets, and pouches.

- **Liquid fertilizer** is a powder or concentrated liquid that's diluted with water and then poured onto soil. Some types are designed to supply nutrients through plant leaves (see "Foliar Feeding" on page 115); these are sprayed onto plants. Fast-acting but more expensive than other types; good for short-term treatment of deficiencies caused by soil alkalinity or acidity.

- **Organic fertilizer** may not actually refer to nutrients from natural animal, plant, or mineral sources. To chemists — and therefore to the fertilizer industry — "organic" usually refers to any compounds that contain carbon. "Organic" may also appear on a blended fertilizer that contains some seaweed or other natural source. If one or more of the N-P-K numbers is higher than 10, a fertilizer has probably been boosted with concentrated, synthetic nutrient sources.

- **Natural fertilizer** usually indicates nutrients from natural animal, plant, or mineral sources. Fertilizer labeling requirements don't give any definition for this term, though. To be sure you know what you're getting, read the ingredient list carefully (and read up on the ingredients later in this chapter).

- **Synthetic fertilizer** is one that contains nutrients formed by manufacturing compounds, as a by-product of industrial processes, or by processing mined minerals with chemicals such as sulfuric acid. Sometimes called "chemical fertilizers," but that's misleading, because natural fertilizers are also made up of chemicals.

Reading the Fine Print

Here's a guide to the fine print on fertilizer labels. Two examples are shown on the next page, since labels vary greatly. The label at the top is from a general-purpose organic fertilizer. The label at the bottom (shown in two parts) is a specialized synthetic fertilizer formulated for acid-loving ornamentals. As you'll see from these examples, the following items don't always appear on labels.

A Guaranteed analysis. Often referred to as the N-P-K, gives nutrient content as a percentage of total weight (pounds per 100 pounds of fertilizer). It's confusing because the percentages don't refer to the elements by themselves. The nitrogen number (8% in the evergreen fertilizer; 4% in the organic) is straightforward; it gives the total amount of that element. But the second number refers to the percentage of phosphate (P_2O_5) available in the first year (4% and 5%, respectively), not the total amount of pure phosphorus (P), which is greater than the listed percentage. The third number refers to potash (K_2O) soluble (available) the first year (4% and 4%), not total potassium (K). (See box at right for conversions.) The nitrogen figure is often broken down to show what forms are present.

B Micronutrients. May be listed on labels of fertilizers that supply them (not required by labeling laws); the "Organic Plant Food" doesn't list them even though the ingredients (poultry manure) will supply some micronutrients.

C Ingredients. Tells the source of each nutrient. Often listed as a breakdown under the guaranteed analysis ("Derived from . . .").

D Filler. The weight remaining after the total of the N-P-K percentages is subtracted. Substances used as filler help in spreading fertilizer evenly or control the rate of release in slow-release fertilizers. In organic fertilizers, fillers may also supply some organic matter or micronutrients. Often not listed (as on these samples).

E Acid equivalent. The amount of acidity this product will cause in soils, expressed in pounds of limestone. Cottonseed meal and several synthetic fertilizers increase soil acidity. Unless you're trying to lower the pH, you'll theoretically need to add the listed amount of limestone to the same area covered by the entire bag of fertilizer to restore the pH. (But remember, don't add limestone and nitrogen fertilizers at the same time or you'll convert the nitrogen to a form that evaporates.)

F Recommended application rates. These amounts are general suggestions for soils that are already in balance. Different rates are usually given for different types of crops/plants. If a soil test shows one or more major deficiencies, you'll have to correct them before following these recommendations.

G Special instructions. For special-purpose fertilizers, labels usually list which plants or types of plants they are designed for, and/or special uses such as foliar feeding.

H Warnings. Label may list precautions for handling, use, or disposal.

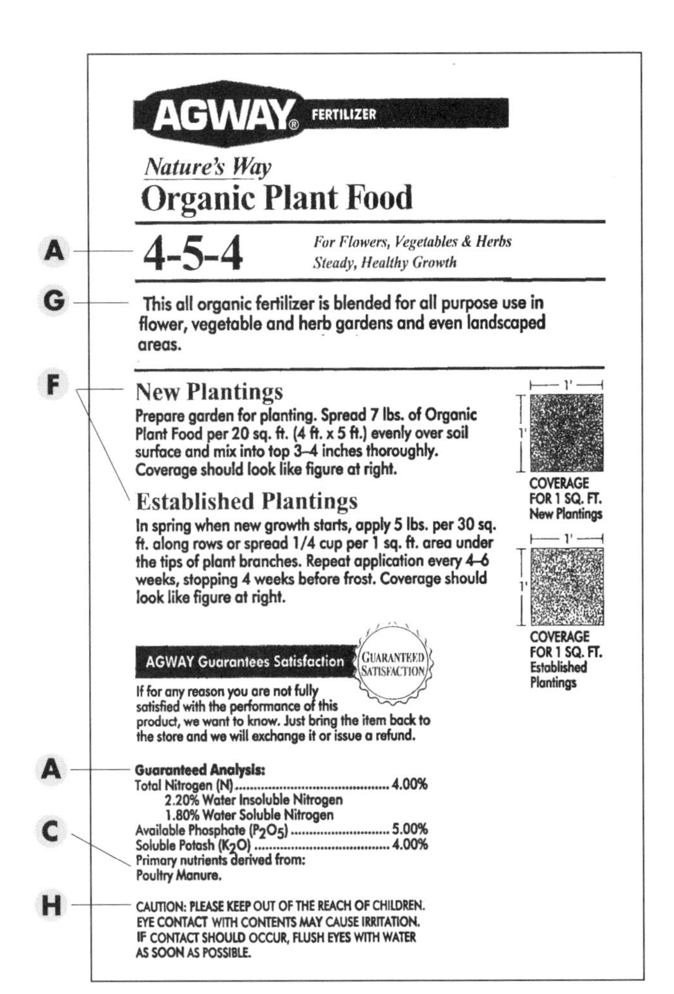

Master Gardening Tip

How Much to Use

Nutrient recommendations are often expressed in pounds per acre. To figure how much fertilizer with a particular analysis to apply, divide the recommended rate by the percentage of that nutrient given in the N-P-K analysis.

For example, to find out how much 12-4-4 fertilizer you need to match a nitrogen recommendation of 45 lbs./acre, divide by 0.12, and then convert to square feet:

45 lbs./acre ÷ .12 = 375 lbs./acre

375 lbs./acre ÷ 43,560 sq. ft./acre x 100 sq. ft. = 0.9 lbs./100 sq. ft.

To find out how much phosphorus (P) is present in a given amount of phosphate (P_2O_5), multiply by 0.44. To translate soluble potash (K_2O) into plain potassium (K), multiply by 0.83. To convert the other way, divide by these numbers instead of multiplying. For other useful conversions, see page 208.

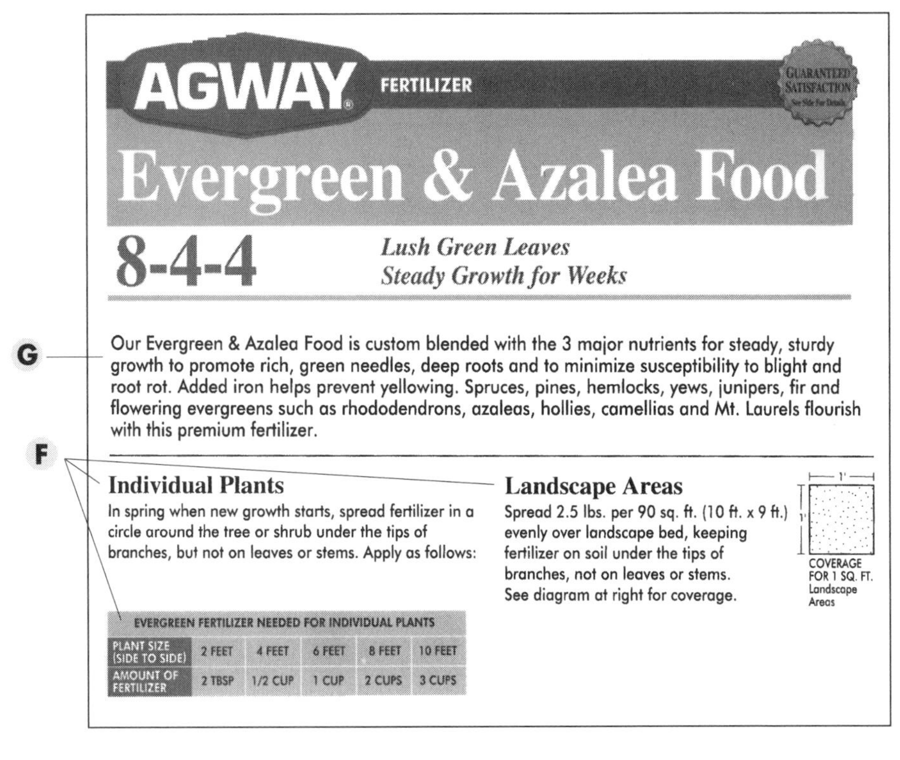

EVERGREEN FERTILIZER NEEDED FOR INDIVIDUAL PLANTS					
PLANT SIZE (SIDE TO SIDE)	2 FEET	4 FEET	6 FEET	8 FEET	10 FEET
AMOUNT OF FERTILIZER	2 TBSP	1/2 CUP	1 CUP	2 CUPS	3 CUPS

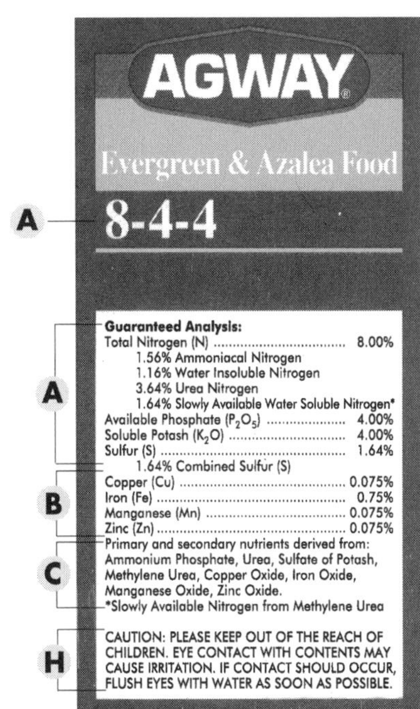

Both labels appear courtesy of Agway Inc.

Ways to Apply Fertilizers

Whatever type of fertilizer you choose, don't exceed the label directions for how much and how often to apply. All-purpose (dry) organic fertilizers are usually applied once in spring, as they release nutrients slowly over the whole season. Apply a little ahead of planting to give them time to start breaking down. Synthetic fertilizers are applied just before planting and may need to be reapplied one or more times during the season. Remember that overfeeding makes plants more susceptible to pests and diseases and can upset the soil balance. When it comes to fertilizers, too little is safer than too much.

Broadcasting. Sprinkle fertilizer evenly over the surface, either by hand or with a spreader. (First measure out the amount recommended for that area.) Rake it in for the fastest response, and to minimize chances of any washing away.

Putting into planting holes. Mix fertilizer well into loosened soil at the bottom of the hole. If using concentrated fertilizers, top with a couple of inches of soil to avoid burning plant roots.

Sidedressing. Spread a band of fertilizer alongside growing plants. Make bands parallel to rows of plants; make the band a ring around the base of a single plant. In either case, keep concentrated fertilizers a few inches away from stems. Scratch the soil lightly to mix in fertilizer.

Watering in. Some fertilizers are made to be dissolved in, or diluted with, water so that you can feed plants as you water them. Some liquid fertilizers can also be used for foliar feeding (see page 115).

Foliar Feeding

Fertilizing plants by spraying or pouring nutrients onto their leaves is called foliar feeding. It works only with liquid fertilizers such as fish emulsion, liquid seaweed formulations, compost or manure tea, and synthetic fertilizers whose labels specify foliar use. Standard fertilizers have to be absorbed by roots before plants can use them.

Follow label directions carefully for dilution and application rates, as products vary. Concentrated products could cause micronutrient imbalances if applied too heavily. Compost and manure teas aren't concentrated and can be applied liberally.

Foliar feeding is the best short-term treatment for nutrient deficiencies. By the time plants show deficiency symptoms, they're already stressed and need help quickly. Absorbing nutrients through their leaves gives plants a boost until soil amendments become available. While the quick fix won't make up for poor soil, it helps you buy time to improve your soil.

Temperature extremes, drought, and even fruit or flower formation can temporarily stress plants. Foliar feeding, especially with kelp or other liquid seaweed products, increases plant resistance to such stress. Used on mature plants late in the season, it can help boost their cold tolerance. Try fish emulsion on seedlings early in the season to supply a quick boost of nitrogen and phosphorus, when cold soil ties up these nutrients. If diluted, it's also good when transplanting.

HINTS FOR SUCCESS

Using Foliar Fertilizers

- Apply foliar fertilizers early or late in the day when it's overcast but not rainy; avoid applying in bright midday sunlight.
- Use either a watering can or a sprayer to apply. With sprayers, adjust the nozzle to give as fine a mist as possible. For more even and thorough coating, add one or two drops of dish detergent to a hand sprayer, or ¼ teaspoon (1.2 ml) to a gallon sprayer.
- Drench both sides of leaves until they're dripping wet; fertilizer that drips onto the soil can be absorbed by roots.

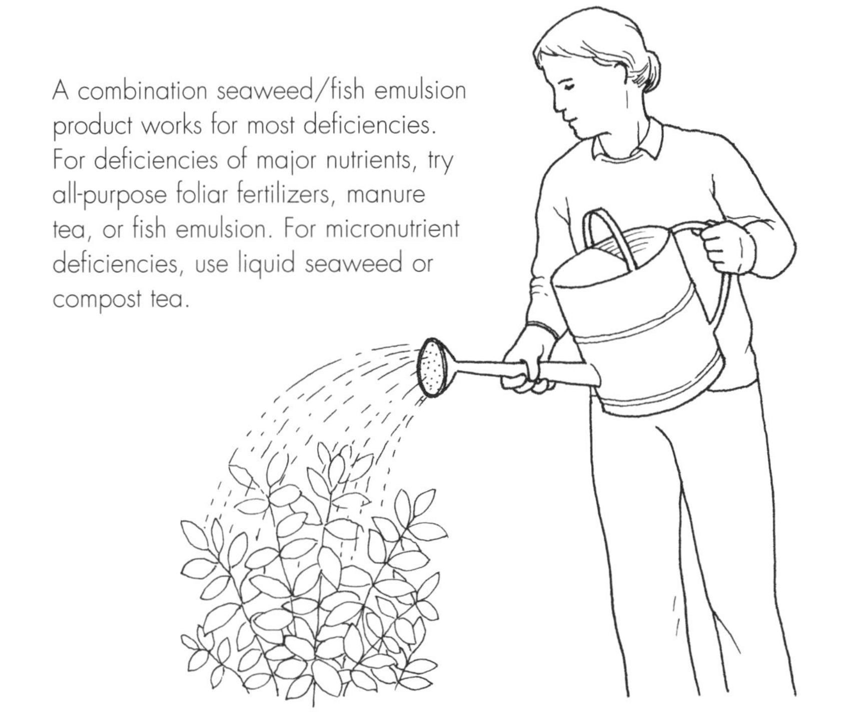

A combination seaweed/fish emulsion product works for most deficiencies. For deficiencies of major nutrients, try all-purpose foliar fertilizers, manure tea, or fish emulsion. For micronutrient deficiencies, use liquid seaweed or compost tea.

Conversions for Liquid Fertilizers

Rate per Acre	Rate per 100 Sq. Ft.
1 pint	¼ tsp. (1.1 ml)
1 quart	½ tsp. (2.2 ml)
1 gallon	1¾ tsp. (8.7 ml)
25 gallons	1 cup (217 ml)

A dilution of 1 pint per 100 gallons = 1 tsp. per gallon = ¼ tsp. per quart.

Supplying Nitrogen

While nitrogen is often considered the most important nutrient in fertilizers, it's also the most difficult to measure in soils. Soil microbes are gatekeepers controlling the day-to-day supply. Extended cold or dry weather (or waterlogging) slows microbes and therefore the amount of nitrogen released to the soil. Since it's harder to calculate how much you need, it's important to find ways to even out the supply.

Even more than for other nutrients, abundant organic matter helps ensure a steady supply of nitrogen — regardless of whether you use organic or synthetic fertilizers. It also cuts down on the amount of fertilizer you need. The recommendations at right are lower than those on many fertilizer labels, which are based on soils with little organic matter.

Using Green Manures and Animal Manure

Fertilizer isn't the only way to supply nitrogen. Using nitrogen-fixing green manures (legumes) plus manure or compost can supply all of a garden's nitrogen needs. On beds where a legume crop was recently turned under, reduce the chart recommendations by half and then be prepared to revise amounts based on your own observations. (Remember that green manures continue to release nitrogen over a long period.)

In addition to the sources given, 12 pounds of dry cow manure (6 pounds of dry poultry manure) per 100 square feet will supply an equivalent amount of nitrogen. You'll need 50 pounds of fresh cow manure, as it's much heavier and less concentrated. Many gardeners prefer to use a combination, such as half this amount with half the recommended amount of dried blood or other source. If you use manure as your only nitrogen source, test periodically for calcium, magnesium, and soluble salts. Where levels of these are already high, manure can create an imbalance. Where these levels aren't high, using manure can reduce the amount of lime needed for acidic soils.

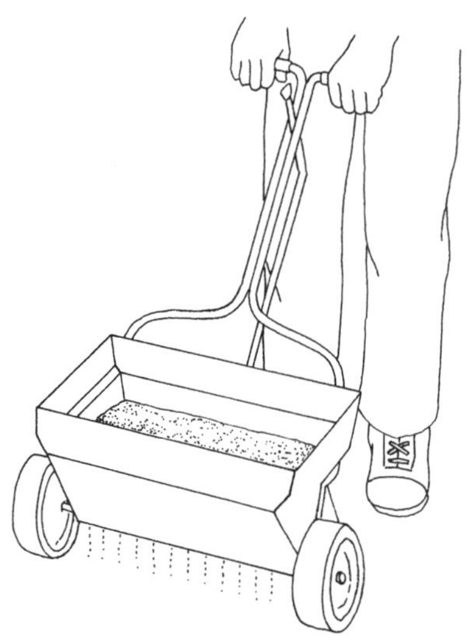

It is the nitrogen in lawn fertilizers that makes lawns green. If you use traditional synthetic fertilizers on your own lawn, avoid creating stripes from uneven coverage by measuring out half the recommended amount and setting the spreader for half the usual rate. Fertilize by walking back and forth in parallel lines. Then spread the remaining fertilizer over the same area, walking at right angles to your original path.

MASTER GARDENING TIPS

Adjusting the Rates

The application rates suggested at right are for vegetable gardens.

- Fast-acting sources of nitrogen wash away quickly, so apply in spring and repeat in midsummer.
- Lawns like lots of nitrogen, so you can apply the recommended amounts twice in one year.
- For flowers, shrubs, and trees in average soil, use half as much, unless they're very heavy feeders (see chapter 8).
- If your soil is fertile, you'll need only half as much to maintain good fertility.
- When a range of formulas is given, the higher application rate is for the lower analysis (weaker formula).
- Note that some nitrogen sources also supply phosphorus and/or potassium. If you use such sources, reduce your applications of the other nutrient(s).
- Avoid overfeeding with nitrogen. Too much causes poor vegetable, fruit, and flower quality (in all but leafy crops) and delays maturity. It increases susceptibility to drought, pests, and diseases, and reduces winter hardiness. Nitrogen is easily washed away, so too much is more apt to pollute nearby groundwater than to build up soil reserves. And before the nitrogen washes away, it could cause a temporary phosphorus deficiency.

Nitrogen Sources

Material	Analysis	Description	Approx. Duration in Soil	Rate for Average Soil (for 100 ft²)	Rate for Deficiency (for 100 ft²)
Natural/Organic Sources					
Blood meal (dried blood)	12-2-0.5	Moderately fast-acting; can burn plants, so keep away from stems; slightly acidifying; repels rabbits and deer but may attract dogs; expensive	3–4 mos.	2 lbs.	3–5 lbs.
Fish meal	5-3-5 to 9-7-2	Supplies major and micronutrients; relatively fast-acting; fishy odor can attract animals; relatively inexpensive	6–8 mos.	2–4 lb.	5–8 lbs.
Fish emulsion	4-1-1 to 9-7-2	Good for foliar feeding; fast-acting; some brands have fishy odor	3–6 mos.	follow directions on label for dilution and application rates	
Tankage	6-6-0 to 9-10-1.5	Dried, ground by-products from slaughterhouse; available from animal feed suppliers	3–6 mos.	2–4 lbs.	5–8 lbs.
Cottonseed meal	6-2.5-1.7	Acidifies soil; contains pesticide residues unless labeled organic (rare); slow-acting	1–2 yrs.	3–4 lbs.	7–8 lbs.
Soybean meal	7-0.5-2.3	By-product from processing soybean oil; slow-acting	1–2 yrs.	3½ lbs.	7 lbs.
Nitrate of soda (sodium nitrate)	16-0-0	Mined form supplies trace elements but synthetic form doesn't; fast-acting; don't use on alkaline or salty soils; about ⅓ as strong as limestone in raising soil pH	3–4 mos.	½ lb.	1 lb.
Synthetic/Inorganic Sources					
Ammonium nitrate	33-0-0	Fast-acting; don't use where soils are high in salt; slightly acidifying; flammable	3–4 mos.	⅓ lb.	¾ lb.
Ammonium sulfate	20-0-0 24% S	Fast-acting; use only on alkaline soils, as it rapidly lowers soil pH	3–4 mos.	½ lb.	1¼ lbs.
Urea	45-0-0	Very fast-acting, high risk of burning plants, so use only in blended fertilizers; mix into soil to minimize loss from evaporating ammonia; increases soil acidity	2–3 mos.	follow directions on label of blended fertilizer	
Urea-formaldehyde	35-0-0	Only moderately fast-acting, so lower risk of burning; safe to use on soils with high salt levels	3–4 mos.	⅓ lb.	¾ lb.
Sulfur-coated urea	36-0-0 13–16% S	Coating makes it slower-acting; lower risk of burning plants; slowly lowers soil pH	1 season	¾ lb.	1½ lbs.

For metric equivalents, see "Useful Conversions" on page 208.

Supplying Phosphorus

Unlike other nutrients, phosphorus doesn't move easily in soil water. It travels at a rate of about an inch a year. That means any phosphorus fertilizers should be applied well before planting and turned under to mix them into the root zone. They can also be scratched into rows before sowing seeds. Soil must be slightly acidic to keep phosphorus from being locked up, so adjust the pH before adding organic or synthetic phosphorus.

Cold temperatures suppress the activity of the soil organisms that make phosphorus available to plants. As a result, young plants growing in cold soils may suffer temporary deficiencies even when soil tests show adequate phosphorus. Where cold spring temperatures last a long time, consider giving early annual crops a short-term boost of bonemeal, composted poultry manure, fish emulsion, or foliar sprays of liquid seaweed.

MASTER GARDENING TIPS

Adjusting the Rates

The application rates suggested in the chart to the right are for vegetable gardens.

- For flowers, shrubs, and trees in average soil you can use half as much, unless they're heavy feeders (see chapter 8).
- If your soil's fertile, you'll need only half as much to maintain good fertility.

Using Rock Phosphate

The easiest way to maintain phosphorus levels over the long term is with rock phosphate (unless you have neutral to alkaline soils). This slow-release form should be spread in fall to be available for the following growing season. It will continue to release phosphorus for at least another three years, so you need to apply it only every four years.

Reading the Labels

Labeling laws require the fertilizer analysis to list the amount of phosphate available the first year. In organic fertilizers, the percentage given in the N-P-K listing is therefore much lower than the total phosphorus present. Bags of rock phosphate list only 3 to 4 percent available phosphate, when the total phosphorus is actually about 30 percent. Common triple superphosphate has a much higher analysis (0-45-0) because all its phosphate is released in the first growing season. Where soils are truly deficient, it's great for bringing up levels quickly. Since it contains less total phosphorus (only 20 percent), it isn't as good a buy as it appears for long-term maintenance of fertile soils.

Protect your lungs and respiratory passages by wearing a dust mask or respirator whenever you handle fine powders such as colloidal rock phosphate.

Phosphorus Sources

Material	Analysis	Description	Approx. Duration in Soil	Rate for Average Soil (for 100 ft^2)	Rate for Deficiency (for 100 ft^2)
Natural/ Organic Sources					
Steamed bonemeal	1-11-0 to 3-30-0 20% P, 24% Ca	Fast-acting; better for alkaline soils than rock phosphates; relatively expensive (raw bonemeal, now hard to find, is slower to break down)	6–12 mos.	1–2 lbs.	2–3 lbs.
Hard rock phosphate	0-3-0 or 0-4-0 30% P, 48% Ca	Very slow-acting; rich in calcium, so not good for most alkaline soils; also supplies iron and micronutrients; relatively inexpensive	5 yrs.	2.5 lbs.	5 lbs.
Colloidal rock phosphate	0-2-0 18% P, 19% Ca	By-product of mining hard rock phosphate; supplies same nutrients but is faster-acting	3–5 yrs.	2.5 lbs.	5 lbs.
Ligno-sulfate rock phosphate	0-3-0 27% P, 29% Ca, 1% S	New product; doesn't require acidic soil to dissolve well; supplies some organic matter and micronutrients	3–5 yrs.	2.5 lbs.	5 lbs.
Synthetic/ Inorganic Sources					
Superphosphate	0-20-0 9% P, 20% Ca, 12% S	Rock phosphate treated with sulfuric acid to make it fast-acting; used in "complete" fertilizers; little risk of burning plants; supplies sulfur and calcium	1 season	1–2 lbs.	2–3 lbs.
TSP/triple superphosphate	0-45-0 20% P, 14% Ca, 1% S	Concentrated form of above but less desirable because it increases soil acidity and supplies almost no sulfur; inexpensive; also called concentrated superphosphate	1 season	½ lb.	1 lb.
Ammonium phosphate	11-48-0 and 18-46-0	Also supplies nitrogen; increases soil acidity; inexpensive; can easily burn plants	1 season	1 lb.	2 lbs.
Nitric phosphate	varies	More accurately called ammonium-nitrate phosphate; fast-acting	1 season	rate varies with formula; follow label directions	

For metric equivalents, see "Useful Conversions" on page 208.

Supplying Potassium

As with nitrogen, a large amount of the soil's available potassium in any given year is taken up by vegetable crops. For other types of gardens, where plants aren't pulled each growing season, only small amounts of potassium are removed from circulation. Most vegetable gardens need some supplemental potassium to replace what's taken up from long-term reserves and to ensure a steady supply to plants. The easiest way to keep soils supplied with potassium is to spread greensand once every four or five years. Other sources are listed in the chart at right.

Avoid Burning and Overuse

Keep synthetic forms of potassium away from the base of plants to reduce danger of burning roots. With the exception of wood ashes, organic sources have a low risk of burning. Fresh organic matter supplies some potassium while increasing long-term potassium reserves.

Some dry-climate soils are naturally well endowed or even oversupplied, so make sure you really need potassium before adding any. As with any nutrient, avoid overusing potassium fertilizers. This nutrient is very easily leached. Before it washes away, excess potassium may cause temporary deficiencies of magnesium and boron. Keeping soils well stocked with organic matter reduces the risk of leaching. Animal manures and most plant residues are also good low-level potassium sources.

Newly transplanted trees, shrubs, perennials, and bulbs benefit from a fall dose of potassium. It increases their cold tolerance and improves their chances for surviving a bad winter. Don't use a source that also contains nitrogen, because a fall dose of nitrogen has just the opposite effect.

MASTER GARDENING TIPS

Adjusting the Rates

The application rates suggested at right are for vegetable gardens and flowers.

- For shrubs and trees in average soil, you can use half as much.
- If your soil's fertile, you'll need only half as much to maintain a good balance.
- If soil tests show low potassium and your soil contains abundant organic matter (or clay), apply the full recommended amount at the beginning of the growing season.
- If you're using greensand, which is slower-acting than other sources, apply it in fall so potassium will be available by the following spring. If you apply greensand in spring, you'll need to supplement it with a fast-acting source, such as a foliar spray of liquid seaweed.
- If your soil is either sandy or low in organic matter and/or clay, it takes more care to guarantee a steady potassium supply. Divide the recommendation into two or three smaller amounts and apply two or three times during the season. This will keep potassium from being leached away before plants can use it. It will also prevent "luxury" consumption. Plants can absorb far more dissolved potassium than they actually need, depleting soil reserves for the next crop.

Potassium Sources

Material	Analysis	Description	Approx. Duration in Soil	Rate for Average Soil (for 100 ft^2)	Rate for Deficiency (for 100 ft^2)
Natural/ Organic Sources					
Greensand (glauconite)	0-1-6 5–7% K	Rich source of micronutrients; loosens clay soils, improves water and nutrient retention in sandy soils; apply in fall, or to compost piles; best for building up long-term reserves rather than treating short-term deficiency; moderately expensive	5–10 yrs.	4–5 lbs.	8–10 lbs.
Sulfate of potash-magnesia (langbeinite)	0-0-22 18% K, 11% Mg, 23% S	Fast-acting, so good for short-term boost; don't use with dolomite limestone or you'll get too much magnesium; also supplies sulfur; doesn't affect pH; good for salty soils; sold as Sul-Po-Mag or K-Mag	3–6 mos.	½ lb.	1 lb.
Wood ashes	0-2-7 fresh, 0-1.2-2 if leached 10–30% Ca, 3–6% Mg, 0.2% Fe	Fast-acting; supplies micronutrients; rich source of calcium, so raises soil pH about ⅔ as much as limestone; don't exceed recommended rates or apply if soil pH is over 6.5; keep dry until spreading to prevent leaching; avoid using on seedlings to prevent burning; not good for acid-loving plants	2 yrs.	1 lb.	3 lbs.
Synthetic/ Inorganic Sources					
Potassium chloride (muriate of potash)	0-0-60 50% K, 47% Cl	Most common source in "complete" synthetic fertilizers; inexpensive; don't use on chlorine-sensitive plants (potatoes, grapes, citrus and other tree crops) or where soil salt levels are high	3–6 mos.	¼ lb.	½ lb.
Potassium nitrate (saltpeter)	13-0-44 37% K	Fast-acting; also supplies nitrogen in a fast-acting but easily leached form; not acidifying; don't use where soil salt levels are high	3–6 mos.	follow label directions to avoid overfertilizing with nitrogen	
Potassium sulfate	0-0-50 42% K, 18% S	No overall acidifying effect on soil; good for chlorine-sensitive plants or where soil chlorine levels are high	3–6 mos.	¼ lb.	½ lb.
Potassium thiosulfate	0-0-25 21% K, 17% S	Reclaimed as manufacturing by-product; used in foliar and regular fertilizers; fast-acting; tends to increase soil acidity	3–6 mos.	½ lb.	1 lb.

For metric equivalents, see "Useful Conversions" on page 208.

Supplying Calcium

Adequate calcium is essential not only for healthy plant growth but also to balance soil chemistry. Calcium is a key that unlocks other soil nutrients; an imbalance of this nutrient can cause toxic levels of some nutrients while tying up others.

Acidic Soils

Soils evolved from limestones start out naturally well endowed. However, calcium is easily leached from soil by water. In order to keep soil from becoming acidic in areas with abundant rainfall (east of the Mississippi and in the Pacific Northwest), gardeners have to add calcium to replenish what gets washed away. (To learn how to correct acidic soils, see pages 150–151.) Gardeners who lime soils regularly won't have to worry about additional calcium. If deficiency symptoms appear in limed soils, look for an environmental cause such as poor drainage. In neutral, reasonably fertile soils, recycling crop residues plus periodically enriching with manure can supply most of the calcium needed.

Once a balanced pH is achieved, test soils every three years or so and amend as needed. If you get your soil tested professionally, follow their recommendations. Lime recommendations supplied with do-it-yourself pH kits vary tremendously and are often too high; use the recommendations in the chart at right instead.

Avoid overliming; an excess of calcium is harder to correct than a deficiency. If you use lots of manure, add less lime. If you apply dolomitic limestone, keep an eye on magnesium levels, as well; too much magnesium can tie up calcium and other nutrients.

Alkaline Soils

Calcium deficiencies can exist in alkaline soils, too. Soil alkalinity can be caused by too much magnesium, potassium, and/or sodium with or without calcium. It's important to determine which elements are causing the alkalinity (with a soil test) in order to understand how to treat it and whether supplemental calcium is needed. Gypsum is excellent for supplying calcium in alkaline soils.

MASTER GARDENING TIPS

"Good" Lime vs. "Bad" Lime

Lime is a general term for all forms of calcium carbonate. The carbonate is even more important than the calcium for raising soil pH.

- Use only agricultural lime (dolomitic or calcitic limestone) in gardens. Builder's lime, also called hydrated or slaked limestone, is too caustic for soils.
- The more finely limestone is ground, the faster it's available. The product label should state that at least 50 percent of the limestone will pass through a 60-mesh screen. Coarser limestone can take several years to break down but it's less expensive; you can use it for maintaining long-term reserves.
- Pelletized limestone works well even if it's larger, as it's mixed with a water-soluble binder that breaks down quickly. It's more expensive but easier to use and less dusty than ground limestone.

Spread lime as evenly as possible. For large areas, use a handheld or push-type spreader. For smaller areas you can sprinkle by hand, but wear gloves to protect your skin. (Details on how to spread lime are given on pages 150–151.)

Calcium Sources

Material	Analysis	Description	Approx. Duration in Soil	Rate for Deficiency (for 100 ft^2)
Natural/Organic Sources				
Dolomitic limestone (dolomite)	25% Ca, 8–20% Mg	Supplies abundant magnesium, so use only if soil tests show need for Mg; slowly raises soil pH, so don't use on alkaline soils	3–4 yrs.	sandy, acid soils: 2–4 lbs. clayey, acid soils: 6–8 lbs.
Calcitic limestone (calcite)	36% Ca	Supplies little or no magnesium; slowly raises soil pH, so don't use on alkaline soils; less expensive than dolomitic limestone	3–4 yrs.	sandy, acid soils: 2–4 lbs. clayey, acid soils: 6–8 lbs.
Crushed oyster shells (ground shell marl)	about 34% Ca	Supplies little or no magnesium; slowly raises soil pH, so don't use on alkaline soils; if finely ground dissolves about as quickly as limestone; coarse grinds are slower-acting but last longer	3–4 yrs.	sandy, acid soils: 2–4 lbs. clayey, acid soils: 6–8 lbs.
Gypsum (calcium sulfate)	22% Ca, 17% S	Also supplies sulfur; can help neutralize overly high magnesium levels; doesn't change soil pH; improves structure of hard, highly alkaline soils; inexpensive	1–2 yrs.	1–4 lbs.
Wood ashes	0-2-7 10–30% Ca, 3–6% Mg, 0.2% Fe	Fast-acting; supplies micronutrients; raises soil pH about ⅔ as much as limestone; rich in potassium so reduce use of fertilizers; keep dry until spreading to prevent leaching; caustic, apply at least 3 weeks before planting; not good for acid-loving plants	2 yrs.	3–4 lbs.
Rock phosphates:	hard: 0-3-0 48% Ca; colloidal: 0-2-0 19% Ca	Very slow-acting — only 2–3% of Ca available in any one year — so good only for building up long-term reserves (colloidal forms are somewhat faster-acting); also supplies slow-release phosphorus and micronutrients; don't use where phosphorus is high	5 yrs.	2–5 lbs.
Synthetic/Inorganic Sources				
Superphosphate	0-20-2 20% Ca, 12% S	Rock phosphate treated with sulfuric acid to make it fast-acting; common in "complete" fertilizers	1 season	use only if phosphorus is deficient (see rates in phosphorus chart)
TSP/triple superphosphate	0-45-0 14% Ca 1% S	Concentrated form of above; lowers pH, so may be useful where alkaline soils are caused by minerals other than calcium	1 season	use only if phosphorus is deficient (see rates in phosphorus chart)

For metric equivalents, see "Useful Conversions" on page 208.

Supplying Sulfur

Soils kept well stocked with organic matter from compost or manure and fertilized with organic sources of nitrogen shouldn't need additional sulfur. If you suspect a sulfur deficiency, check with your local Cooperative Extension Service to see if soils in your region tend to have unusually low sulfur levels. If you know that triple superphosphate, ammonium nitrate, or urea fertilizers have been used on your soils for some time, you should probably get a soil test for sulfur deficiency.

Even though sulfur is used to lower soil pH, it can still be deficient in soils that are acidic. It's slowly released from organic matter by microbial activity, but it can be leached away. If your soil is low in organic matter and acidic from leaching by rainwater, get busy replenishing organic matter levels to supply sulfur.

The chart below describes the options available for adding sulfur if tests show deficiency. To learn how to use sulfur to correct alkaline soils, or gypsum to treat salty soils, see chapter 7.

Sulfur Sources

Material	Analysis	Description	Approx. Duration in Soil	Rate for Deficiency (for 100 ft^2)
Natural/ Organic Sources				
Sulfate of potash magnesia (langbeinite)	0-0-22 23% S, 11% Mg	Fast-acting, so good for short-term boost; don't use with dolomite limestone or you'll get too much magnesium; doesn't affect soil pH; sold as Sul-Po-Mag or K-Mag; somewhat expensive	3–6 mos.	½–1 lb.
Gypsum (calcium sulfate)	22% Ca, 17% S	Also supplies calcium; can help neutralize overly high magnesium levels; doesn't change soil pH; improves structure of hard, highly alkaline soils; inexpensive	1–2 yrs.	1–4 lbs.
Elemental (agricultural) sulfur	88–99% S	Releases sulfur slowly, but lowers soil pH quickly, so use only on alkaline soils; add in fall with manure, compost, or other organic matter and mix into top 3 inches of soil	1–2 yrs.	1 lb.
Epsom salts (magnesium sulfate)	13% S, 10% Mg	Fast-acting, so apply in spring; also supplies magnesium; can use as a foliar fertilizer; relatively expensive	1–2 yrs.	½ lb.
Synthetic/ Inorganic Sources				
Superphosphate	0-20-0 9% S, 20% Ca	Fast-acting; common in "complete" fertilizers; safe to use on salty soils; does not affect soil pH; use only if calcium and phosphorus are deficient	1 season	1–2 lbs.
Ammonium sulfate	20-0-0 24% S	Fast-acting; avoid using where soil salt levels are high; increases acidity of soil; concentrated source of nitrogen, so don't add any other nitrogen	3–4 mos.	½–1 lb.
Potassium sulfate	0-0-50 18% S	Fast-acting; good source of potassium for soils with high chlorine content; does not affect soil pH; use only if potassium is deficient	3–4 mos.	½–1 lb.
Potassium thiosulfate	0-0-25 17% S	Used in foliar and regular fertilizers; fast-acting; tends to increase soil acidity; use only if potassium is deficient	3–4 mos.	½–1 lb.

For metric equivalents, see "Useful Conversions" on page 208.

Supplying Magnesium

Many soils are naturally well endowed with magnesium, which is why it's not included in most fertilizers. The soils most apt to be deficient are sandy ones. These soils also tend to be acidic, though, and the most common treatment for soil acidity — dolomitic limestone — contains enough magnesium to correct most deficiencies.

If magnesium levels are low but calcium levels are fine, don't add dolomitic limestone. Instead, use sulfate of potash-magnesia if potassium levels are also low. If both calcium and potassium levels are adequate or high, you'll need to use Epsom salts to supply magnesium.

Excessive amounts of calcium or potassium in the soil can lock up magnesium and create a temporary deficiency. Adding more magnesium isn't the answer. Either of these conditions causes the soil to be too alkaline. Correcting the alkalinity by adding sulfur is the best way to fix this temporary magnesium deficiency.

An excess of this essential nutrient will lock up potassium, zinc, boron, and manganese. It can also interfere with soil structure, making it sticky. Gardeners with poor soil structure or drainage problems should be on the lookout for too much magnesium, as it will just compound their problems.

Use Epsom salts in a foliar spray to get the most benefit for the least cost. Mix ¼ pound (113 g) in 2½ gallons (9.5 l) of water and spray on leaves several times a season.

Magnesium Sources

Material	Analysis	Description	Approx. Duration in Soil	Rate for Deficiency (for 100 ft²)
Natural/ Organic Sources				
Dolomitic limestone (dolomite)	8–20% Mg, 25% Ca	Most common source; raises soil pH, so don't use on alkaline soils	3–4 yrs.	sandy, acid soils: 2–4 lbs. clayey, acid soils: 6–8 lbs.
Epsom salts (magnesium sulfate)	10% Mg, 13% S	Fast-acting, so apply in spring; use only on soils that already have high levels of potassium and calcium; no effect on soil pH	1–2 yrs.	½ lb.
Sulfate of potash-magnesia (langbeinite)	0-0-22 11% Mg, 23% S	Fast-acting, so good for short-term boost; don't use with dolomite limestone or you'll get too much magnesium; doesn't affect soil pH; good for salty soils; sold as Sul-Po-Mag or K-Mag; somewhat expensive	3–6 mos.	½–1 lb.

For metric equivalents, see "Useful Conversions" on page 208.

Supplying Micronutrients

For most gardens, correcting soil pH and spreading compost or manure every year or two will maintain adequate levels of all micronutrients. Micronutrient deficiencies are most often a problem in sandy, acidic soils in areas with abundant rainfall, or on very alkaline soils. Soils that have been intensively farmed or gardened, where only "complete" synthetic fertilizers have been used without abundant compost or other organic matter, are also apt to be deficient.

Specific deficiencies can be treated with liquid seaweed or chelates of the element, either applied as a foliar spray or added to soil. Since chelates are relatively expensive, they're used mostly for short-term treatment of ornamentals and fruit trees. Chelates are complex molecules that hold nutrients loosely until plants need them.

Several micronutrient fertilizers and blends are produced as frits. These are glasslike beads that bind micronutrients and release them slowly as beads break down, keeping even alkaline soils from locking them up. Other fertilizers for treating deficiencies are listed below. Since the amounts needed to treat deficiencies are so low (only .0009–0.3 lbs./acre!), and since only slightly larger amounts can kill sensitive plants, be careful. Consult your Cooperative Extension Service or other professional before using chelates or other sources of individual micronutrients. Follow label directions carefully.

Intensely cultivated vegetable gardens may benefit from a balanced general micronutrient boost every few years. Seaweed and granite dust are both good for this. Rock phosphate and greensand also supply many micronutrients, as well as phosphorus and potassium (respectively). Any of these can be added to the compost pile every few years to ensure a good balance. Some people prefer to add a little directly to their vegetable gardens every four or five years (mixing with compost before spreading gives a better distribution). See the fertilizer recipes at the end of this chapter for blends that supply micronutrients.

Fertilizers for Specific Deficiencies

Micronutrient	Plants Most Likely to Show Deficiencies	Source	Amount of Nutrient Supplied
Boron	Spinach, turnip and other cabbage-family crops, asparagus, beet, celery, peanut, apple, pear, grape, alfalfa, clover, pine and other conifers	Borax (sodium borate)	11%
		Sodium pentaborate	18%
		Boron frits	2–11%
Copper	Spinach, beet, onion, carrot, lettuce, corn, citrus and other large fruits, wheat and other grains (and grasses)	Copper sulfate	26% (13% S)
		Copper chelates	9–13%
		Copper oxides	75%
Manganese	Spinach, beet, peas and beans (legumes), potato, onion, lettuce, radish, many fruits, oats, wheat	Manganese sulfate	25–28%
		Manganese oxide	41–72%
Molybdenum	Spinach, beet, beans and other legumes, broccoli and other cabbage-family crops, lettuce, tomato	Sodium molybdate	39%
		Ammonium molybdate	54%
		Molybdenum trioxide	66%
Zinc	Onion, beans and most legumes except peas, tomato, corn, rice, grape, most fruit and nut trees, pines	Zinc sulfate	23–35%
		Zinc oxide	78–80%
		Zinc chelates	9–14%

Supplying Iron

Chlorosis — yellowing between leaf veins — is an easy-to-recognize condition that usually indicates iron deficiency. Acid-loving plants such as azaleas, rhododendrons, and blueberries commonly show iron chlorosis. Other plants that are particularly susceptible include roses, hollies, grapes, beans, and several types of shade trees. Iron is often present but unavailable in alkaline soils and in soils with too much calcium, phosphorus, copper, zinc, or manganese. Acid-loving plants can show deficiencies even in neutral to slightly acidic soils.

Correcting the pH and maintaining adequate levels of organic matter will cure most cases of iron deficiency. If changing the pH doesn't correct the problem, have your soil tested for iron. Ask your local Cooperative Extension Service if soils in your region tend to be iron-poor. Ferrous sulfate (19 percent Fe) is the most common synthetic fertilizer used to supply iron to soils that are lacking.

MASTER GARDENING TIP

Using Liquid Seaweed and Chelates

For short-term treatment of mild chlorosis, try spraying plants with (diluted) liquid seaweed. For more severe cases, use iron chelates, which contain about 6 percent iron. Since humus is a natural chelate, it's excellent for increasing the availability of iron and other micronutrients. While commercial iron chelates are relatively expensive, they can restore the healthy, green color to leaves while you're waiting for soil amendments to bring the pH into balance. Foliar chelate sprays work in a few days; soil applications take longer. Follow label directions carefully.

MASTER GARDENING TIPS

Using Rock Dusts

Rock dusts (also known as rock powders) are an excellent way to supply micronutrients (also beneficial-but-not-essential minerals). They're very slow-acting but release nutrients only as they're needed.

- ***Limestone*** is the best-known rock dust. Since it also supplies large amounts of one or more major nutrients, it's not usually included when gardeners talk about rock dusts.
- ***Greensand*** (0-1-6) contains enough potassium to be considered a good source of that nutrient. It also contains up to 50 percent silica (sand) and 32 other minerals.
- ***Granite dust*** (also called granite meal) contains only half as much potassium, 19 minerals, plus about 67 percent silica. While not an essential nutrient, silica improves soil texture and increases disease resistance.

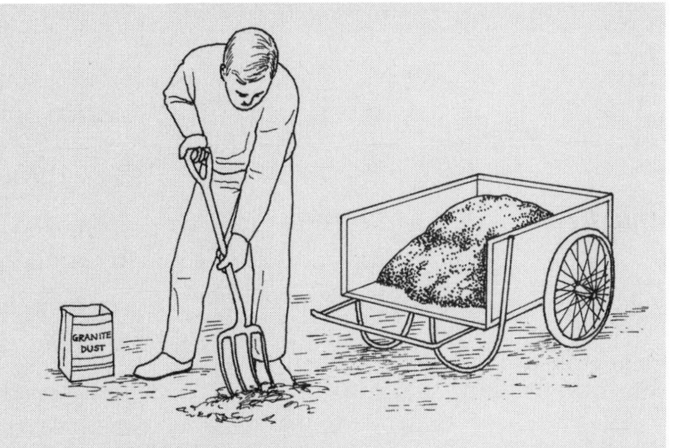

The smaller the rock dust particles, the sooner the nutrients will be released. Check labels to make sure rock dusts are fine enough to pass through a 200-mesh screen. Mix thoroughly into soil and add organic matter at the same time. Some people prefer to add rock dusts (other than limestone) when building a compost pile.

Using Animal Manures

Manure is the most ancient fertilizer, and still one of the best. In addition to supplying low levels of major nutrients, all manures are rich sources of calcium, micronutrients, and organic matter. Manure is also rich in beneficial microorganisms, making it an excellent addition to compost piles as well as to garden soil. The nutrient content of manure varies widely, so the values given here are only approximate.

Fresh manure. Fresh manure has the highest levels of nitrogen, microorganisms, and weed seeds. The nitrogen (especially in chicken manure) can burn plant roots and interfere with seed germination. Avoid spreading fresh manure around plants unless they're dormant. You can mix in a 2- to 4-inch layer at the end of the fall growing season to prepare beds for spring planting. For fresh chicken manure, use half as much and cover with an equally thick layer of carbon-rich shredded leaves or straw. Composting fresh manure will prevent burning and reduce weed seeds. Never use "manure" from dogs or cats, as these can carry human diseases.

Dried and composted manures. Dried manures, available in easy-to-handle bags at garden centers, are less apt to burn plants. They're a more concentrated nutrient source because the water has been removed. For most dried manure, apply 15 pounds per 100 square feet (7.3 kg/10 m^2). For stronger types — sheep, poultry/chicken manure, bat guano, and cricket castings — use half as much. Many suppliers sell bagged, composted manures. These have no odor and won't burn or introduce weed seeds. Apply them at the same or slightly higher rates than dried manures; check labels for recommended application rates.

Types of Animal Manures

Type	Average Nutrient Content (approx.)	Comments
Cow, fresh	0.6-0.3-0.6 3% Ca, 0.1% Mg	Few weed seeds; strong odor; slower to decay than other manures
Cow, dried	2-2.3-2.4 2.9% Ca, 0.7% Mg	Won't burn plants; few weed seeds; use sparingly where soil salt levels are high
Horse, fresh	0.7-0.3-0.6 0.3% Ca, 0.1% Mg	Can burn plants; may contain lots of weed seeds; mild odor
Sheep/goat, fresh	1.2-0.4-1.0 0.2% Ca, 0.3% Mg	Pellets are easy to handle; few weed seeds; little odor; contains more organic matter than other fresh manures
Pig, fresh	0.5-0.3-0.5 0.2% Ca, 0.03% Mg	Strong odor; high moisture content makes it slower to decay than other manures
Chicken, fresh	1.3-2.7-1.4	High in available nitrogen; will burn plants unless composted first; wear respirator to protect lungs from fungus; strong odor
Poultry, fresh	1.1-0.8-0.5 0.4% Ca, 0.2% Mg	Often contains chicken manure, but mixed with turkey, duck, or other domesticated bird droppings
Dried chicken/ poultry	4.5-2.7-1.4 2.9% Ca, 0.6% Mg	Good for boosting phosphorus in cold or alkaline soils (add with organic matter); wear respirator to protect lungs from fungus
Earthworm castings	0.5-0.5-0.3	No risk of burning plants; superb soil conditioner; can use up to 25 lbs. per 100 sq. ft. for poor soils; very high in organic matter
Bat guano	8-4-2	Dry bat droppings collected from caves, so nitrogen hasn't leached away
Cricket castings	4-3-2	From crickets raised as fish bait; rich source of calcium and micronutrients

Making Manure Tea

Manure tea is a great liquid fertilizer that's easy to make. Use it full strength around plants for a quick nutrient boost, as when the soil is cold in spring. Or, instead of topdressing with composted manure, use it for periodic feedings. Dilute before using on plant leaves, though; they don't like the strong stuff. You can use the diluted tea as often as you want to water fast-growing, heavy-feeder vegetables such as broccoli and salad greens. To dilute, mix with 3 to 4 parts water until it's the color of weak tea.

1 Place a large shovelful of manure into a burlap bag and close the top of it securely. Place the bag in a large bucket and add water. Swirl your "tea bag" around, making sure it soaks up enough water to keep it submerged. Let your liquid fertilizer steep for a week.

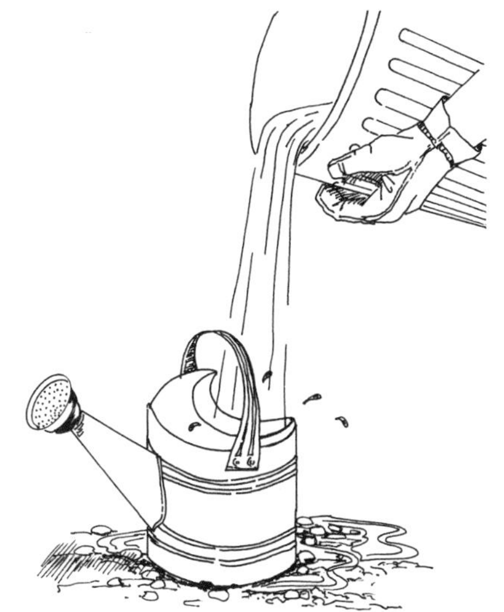

2 Pour off the finished tea into a watering can. Use full strength on soil or dilute to one-quarter strength for foliar feeding. Dump the contents of your "tea bag" onto the manure pile.

MASTER GARDENING TIP

Making Compost Tea

Compost tea is made the same way as manure tea, using compost instead of manure. Fresh, finished homemade compost made with some manure is best, but commercial types also work. Good compost tea doesn't smell so nice, because it's fermented. If you can handle the smell, allow it to brew twice as long as manure tea. You can shovel compost directly into the bucket, and strain the finished product. Before using on plants, dilute with about an equal volume of water. It should be the color of weak tea.

While compost tea supplies only low levels of major nutrients, it's good for a micronutrient boost. Plants stressed by drought or insects are particularly good at soaking up its nutrients through their leaves. Drenching plant leaves with it is one of the best controls for fungal leaf-spot diseases. Compost tea can help prevent as well as control such diseases.

For the best results, use the tea within three weeks of brewing. Remove badly diseased leaves before drenching plants. Early in the evening or on a cloudy day, pour tea over plants (lifting leaves if necessary) until upper and lower sides of leaves are dripping wet. Repeat in two or three weeks. You can continue applying every two to three weeks for as long as you like. Plants can't overdose on compost tea.

Natural Fertilizers

Many organic materials can be used as fertilizers. All provide micronutrients. While levels of major nutrients may seem small, they're released over a long time. Nutrient contents vary. Those listed here are only averages; the nutrient analysis listed on product labels is more accurate. The application rates recommended here are for average soils (those with no major deficiencies), to maintain adequate nutrient levels. If your soil is fertile, use half as much; if it's low in nutrients, use half again as much.

Type	Average Analysis (approx.)	Amount to Apply (for 100 ft^2)	Comments
Alfalfa:			
meal	3-1-2	3½ lbs.	Good source of organic matter, nutrients, and a natural plant-growth stimulant (triacontatol); good soil conditioner
hay	2.5-0.5-2	15 lbs.	
Coffee grounds	2-0.3-0.3	—	Use in compost or in mulch; acidic, so add a little limestone or wood ashes to compensate
Commercial compost	1-1-1	10 lbs.	Excellent source of organic matter and micronutrients; great soil conditioner; available at most garden centers
Eggshells	1.2-0.4-0.1	—	Excellent source of calcium and micronutrients; dry overnight and crumble onto soil or into compost pile
Fresh grass clippings	0.5-0.2-0.5	—	Incorporate into soil or add thin layers to compost piles; avoid thick layers, which form sticky mats that repel water
Grass hay	1.2-0.4-1.5	15 lbs.	Good source of organic matter, but lower in nitrogen than alfalfa hay
Kelp meal	1.5-0.5-2.5	1–2 lbs.	Excellent soil conditioner; supplies micronutrients and increases plant resistance to stress
Lobster shells	4.6-3.5-0	—	Can attract animals; crush to speed decomposition
Milk	0.5-0.3-0.2	—	Where animals aren't a problem, add soured milk to compost piles
Oak leaves	0.8-0.4-0.2	15 lbs.	Chop before using; add to soil in fall for spring gardens, use as mulch, or compost; more acidic than other leaves (all leaves work well)
Peanut shells	0.8-0.2-0.5	—	Can help aerate compost piles; high in carbon, so nutrients will be available more quickly if a nitrogen source is added at the same time
Pine needles	0.5-0.1-0	—	Acidic, so use with wood ashes or limestone on most plants; excellent mulch for acid-loving plants
Salt-marsh hay	1.1-0.3-0.8	15 lbs.	Available in coastal areas; good for mulch
Sawdust	0.2-0-0.2	15 lbs.	Compost first, or mix with nitrogen source such as blood meal (3 lbs. per bushel) or lots of coffee grounds before adding to soil
Straw	0.6-0.2-1.1	15 lbs.	High carbon content, so nutrients will be available more quickly if a nitrogen source is added at the same time

For metric equivalents, see "Useful Conversions" on page 208.

Mixing Your Own Balanced Fertilizers

In addition to providing nitrogen, phosphorus, and potassium in slow-release form, a balanced fertilizer should supply smaller amounts of calcium, magnesium, sulfur, and micronutrients to maintain good tilth. A mix of organic amendments is the best way to meet these requirements and to maintain long-term fertility. (Remember to test pH periodically and adjust as needed so that all nutrients remain available.) Blended organic fertilizers are available at garden centers. If you want to mix your own, here are some easy recipes. *(For metric equivalents, see "Useful Conversions" on page 208.)*

Moderately Fast-Acting 5-10-10 Fertilizer (for acidic soils)

1–2 bushels well-rotted manure
1 pound blood meal
6 pounds steamed bonemeal *or*
16 pounds rock phosphate
8 pounds wood ashes (alkaline soils: use 5 lbs. granite dust)

Yield: about 25 pounds

This formula is good for most vegetables and ornamental foliage plants. It also supplies organic matter. It makes enough for 100 square feet on average soil; use only half this recipe for fertile soil. The ingredients are moderately fast-acting for use at the beginning of the growing season (there's enough rock phosphate to cover the first season, even though it's slow-acting). Since it contains lots of wood ashes, it should be used only on soils with a pH of 6 or lower and replaces the need for limestone. On more alkaline soils, substitute a mica-rich type of (slower-acting) granite dust for the wood ashes and don't use rock phosphate. You'll need a large scale to make this recipe, which is blended by weight.

Moderately Fast-Acting 5-10-5 Fertilizer

2 pounds blood meal
6 pounds bonemeal
8 pounds granite dust
4 pounds wood ashes

Yield: 20 pounds

This general-purpose formula is high in phosphorus. As a result, it's particularly good for flowers and vegetables such as squash and melons. The amount of wood ashes is low, so you can use it on all but alkaline soils. It doesn't supply organic matter; use with compost, lots of mulch, or other humus source. The ingredients break down relatively quickly, so apply at the beginning of the growing season. Use 3 to 4 pounds for 100 square feet. It's also blended by weight, so you'll need a large scale.

Slow-Release Soil Sustainer

3 parts fish meal
6 parts hard rock phosphate *or* colloidal rock phosphate
1 part kelp meal *or*
6 parts greensand

Yield: either 10 or 15 parts

This general-purpose formula works for all types of plants. The first ingredients make a roughly 4-6-3 fertilizer; substituting one or both of the alternate ingredients will lower the analysis somewhat. Try a cup (small garden) or a coffee can (medium garden) to measure parts. As these amendments take longer to release nutrients, apply this formula in fall to supply nutrients for the following spring. Additional phosphorus and potassium will become available in the following years as a result of microbial activity. Apply 3 to 5 pounds for each 100 square feet (use the higher rate if you've used one or both of the substitute ingredients). Use along with compost or another source of organic matter.

Growth-Enhancing Substances

Gardening magazines carry ads for mysterious substances that promise miracles in gardens. Reputable catalogs offer clearly identified soil additives that promise varied benefits. Just what are these products?

Each may contain one or more substances that act in different ways, so it's easiest to lump them together as growth enhancers. While fertilizers also promote growth, these products are different (even though some also contain micronutrients). They provide the most striking benefits to soils of low fertility, or where synthetic fertilizers have been used for years.

- **Chelating agents.** These naturally occurring compounds help keep soil nutrients available, especially micronutrients. They stop them from washing away or from being converted into insoluble forms. Treatments for chlorosis usually contain iron plus chelating agents to help keep the iron available. Humus is a chelating agent (among other things). If your soil's rich in organic matter, you won't need to add extra chelating agents.

- **Enzymes.** In soil, some enzymes (biochemical activators) can help release nutrients or convert them into forms that plants use. Others increase wettability of soil, improving water absorption. Soils rich in organic matter (especially compost) and fertilized with organic amendments don't need extra enzymes. Synthetic fertilizers can interfere with enzyme activity, so wait a week or two after using them to apply products containing enzymes.

- **Humic acid.** This is one of the compounds in humus. It's dark brown and causes permanent stains. You can assume it's in any liquid growth promoter that's dark and warns about staining. Humic acid provides the same benefits as humus; you don't need to add it to soils already rich in well-decomposed organic matter.

- **Plant hormones** (plant growth substances). Plants produce these substances to stimulate different types of growth and development, from cell division to the formation of roots or flowers. Auxins, cytokinins, and gibberellins are the main hormones. These often don't appear on labels of growth-enhancing formulas because they're usually supplied by marine algae or seaweed. Alfalfa contains another plant growth substance.

MASTER GARDENING TIP

Kelp and Seaweed Fertilizers

Kelp is a particularly nutrient-rich form of seaweed. All seaweed products are good for supplying major and micronutrients, but kelp seems to provide even more benefits. It enhances plant growth, even though the exact compounds and mechanisms responsible aren't understood. It supplies over 40 nonessential minerals; some of these may contribute to its benefits. It can even help improve soil structure. Kelp increases frost resistance, perhaps simply by making plants healthier. Kelp also increases resistance to pests and diseases.

You can buy kelp as dried meal or in liquid form. Add kelp meal to soil in fall (2 to 3 pounds per 100 square feet) for benefits the following year. For faster results, drench plants with liquid kelp monthly throughout the growing season.

Liquid seaweed products vary. Not all contain kelp, and even those that do will vary in nutrient content. The more concentrated formulas are also good for supplying fast-acting major nutrients. Dilute according to label directions, as they can burn leaves if used full strength.

CHAPTER · 7

Improving New Sites and Problem Soils

Whether you're starting a new garden or improving an existing one, the approach is the same: Enhance the good qualities of your soil and learn to manage the less-than-ideal aspects. Fortunately, you need to know only a few techniques for any soil.

Don't Work So Hard!

There's good news for people whose gardening time and energy are limited. Evidence shows that less is better when it comes to tilling. Repeated tilling or plowing can create a layer of hardpan just below the reach of the tines. Even infrequent tilling breaks down soil structure. On heavy or clay soils mechanical tilling or hand digging of wet soil can create bricklike clods and impair drainage.

Try to turn over your soil no more than once or twice a year. Fall is a great time to mix in amendments and organic matter, which you can do in a single tilling. In spring simply rake in a bit of fertilizer as you smooth the soil surface. If you're growing winter cover crops, skip the fall tilling and mix in any additional amendments when you till in spring.

In ornamental plantings, skip tilling entirely. Use mulch for weed control and pull the mulch aside for any supplemental feedings. Scratch fertilizer into the surface with a hand cultivator before replacing the mulch.

In every type of garden, let earthworms do most of your work for you. When they dig, they not only mix the soil, but also open up drainage channels and build up good structure.

In This Chapter

Starting a New Garden

You can dig new beds at any time of year, but fall offers distinct advantages. In most areas, soils tend to be drier then. You can use "raw" sources such as chopped leaves, sawdust, and unfinished compost for your organic matter; they'll have plenty of time to compost in place (and release some nitrogen) by spring. Rock powders (see page 127) are superb nutrient sources because they don't overload the soil, but they also require time to work. Finally, the window of opportunity is longer in fall than in spring. You can choose any fall day before the ground freezes, preferably one of those gorgeous, crisp days when it's a joy to work outdoors. In spring, the period is shorter between when the ground thaws or dries out and planting time.

DID YOU KNOW?

Reducing Waists

Did you know that shoveling and digging burn 200 to 300 calories per half hour? Turning compost piles burns a similar amount of calories. They're good overall workouts, including all of the major muscle groups (legs, stomach, arms, shoulders, back, and neck). Forty-five minutes of gardening burns as many calories as a 30-minute aerobics session. If you're out of shape, take breaks and straighten up frequently to prevent back strain.

Single-Digging

Simple cultivation, following the basic method outlined here, improves your soil and allows you to grow most plants in most conditions. It works best for loamy to sandy soils. If you suspect heavy or clay soils, or drainage problems, turn to later sections of this chapter to learn the advantages of double-digging and raised beds. If you don't know what kind of soil you have, use the simple tests in chapters 1 and 2 to find out.

Try to minimize walking on your beds once you've invested effort to prepare them. Create paths with stepping-stones, mulch, scrap pieces of wide boards, or even strips of old carpet. *(For metric equivalents, see "Useful Conversions" on page 208.)*

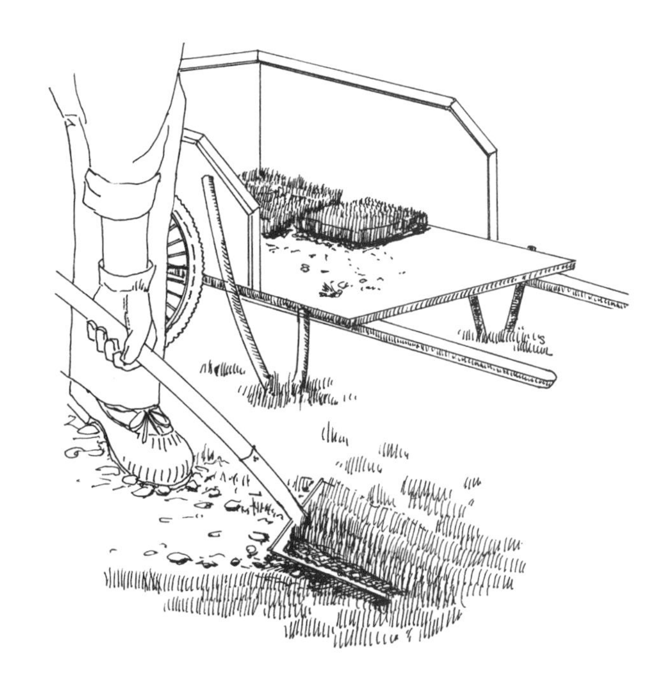

1 Remove existing sod with a sharp-edged spade: Slice around the edges of a 1-foot-square piece and slide spade under roots about 3 inches deep to lift off sod. Lay sod upside down and one layer thick in your compost pile.

2 Spread a 2- to 3-inch layer of organic matter evenly over the area. Also spread powdered rock phosphate and greensand (follow label recommendations or those in chapter 6). Test your soil's pH before spreading lime or sulfur to correct acidity or alkalinity.

3 With a spade or garden fork, dig soil to the depth of the blade or tines. Turn over each spade- or forkful in place. Continue turning until no large clumps of soil or organic matter remain. Or use a rotary tiller to turn the top 6 to 8 inches. If soil will remain bare for over a month, cover with a coarse layer of mulch.

4 Just before planting, rake any added fertilizers into the top few inches of soil and then rake smooth.

Master Gardening Tips

The Green Manure Alternative

If you can plan ahead an entire season, growing green manures offers two distinct advantages over basic bed preparation. First, you can add large amounts of organic matter without importing or hauling materials or making compost. You simply grow all the material right where it's needed and then turn it under. Second, you eliminate many weed seeds, so you'll greatly reduce the time spent weeding your future garden. (See chapter 6 for specific instructions.)

The No-Till Alternative: Sheet Composting

If you don't want to dig or remove sod, use the sheet-composting steps explained on pages 76–77. You'll need more time (six months) but less overall effort.

Managing Problem Soils

The basic strategy for problem soils is simple: Fix what you can and learn to manage what you can't fix. Moderate cases of alkalinity or acidity are good examples of characteristics you can fix. Using muscle power, you can also fix some drainage problems by physically breaking up or cutting holes through hardpan. This chart outlines strategies for specific problems; read on in this chapter for details.

Save most of your effort for those few plants you just have to grow that can't tolerate existing conditions. With really problematic soils, try growing plants in containers or excavating a few inches of soil and filling in with imported topsoil to create raised beds. It's often easier to go with the flow: Design a garden around plants that thrive in the conditions your yard offers. (Ask your Cooperative Extension Service, local nurseries, or local landscapers.) You'll be surprised at the variety you have to work with.

Above all, remember that organic matter is the universal soil improver (see chapters 3 and 4). That's why adding it is listed as the top strategy for each entry in the chart.

Quick Reference Chart

Problem	Management Strategies	Where to Find Reference
Soggy soils or poor drainage (see also clay soils and compacted soils)	• Add organic matter (include some relatively fresh residues)	chapters 4 and 5
	• Identify cause (topography or soil structure)	pages 10–12
	• Regrade low spots; loosen compacted soil with broadfork	pages 138, 143
	• Double-dig beds or build raised beds	pages 138–139, 141
	• Improve and protect soil structure	page 10; chapters 4, 5
	• Check for extreme magnesium levels	pages 140, 206
	• Choose plants that tolerate soggy soils	pages 138, 183
Clay and heavy soils (see also poor drainage)	• Add organic matter (include some relatively fresh residues)	chapters 4 and 5
	• Double-dig beds or build raised beds	pages 138–139, 141
	• Improve and protect soil structure	page 10; chapters 4, 5
	• Avoid overwatering and compacting soil or working when wet	pages 10, 142
	• Choose plants that tolerate heavy soils	page 140
Compacted soils or hardpan (see also poor drainage)	• Add organic matter (include some relatively fresh residues)	chapters 4 and 5
	• Loosen soil with broadfork and/or double-dig beds (or cut holes through hardpan)	pages 138, 143
	• Improve and protect soil structure	page 10; chapters 4, 5
	• Grow deep-rooted green manures	chapter 5
Sandy soils or excessive drainage	• Add organic matter (well-decomposed is best)	chapters 4 and 5
	• Create sunken beds (where salt levels are low)	page 145
	• Mulch to reduce evaporation	page 144; chapter 5
	• Use fertilizer often, but in small amounts (or use foliar fertilizer)	chapter 6
	• Choose drought-tolerant plants	pages 144, 183

Problem	Management Strategies	Where to Find Reference
Steep slopes (see also excessive drainage)	• Add organic matter (well-decomposed is best) • Plant sturdy groundcovers or build terraced beds • Minimize cultivation • Cover any exposed soil with mulch (or grow cover crops)	chapters 4 and 5 pages 146–149 page 147 chapter 5
Thin and/or stony soils	• Add organic matter • Build raised beds or grow plants in containers • Choose plants adapted to local area	chapters 4 and 5 pages 138–139, 168–169 pages 155, 181
Acidic soils	• Add organic matter to increase buffering • Test soil to determine if calcium and/or magnesium is needed • Spread limestone and retest after a year • Switch to fertilizers and amendments that don't acidify soil • Choose plants that tolerate acidic soil	chapters 4 and 5 pages 122, 125, 150 pages 150–151 chapter 6 pages 150, 182
Alkaline soils, also saline and sodic soils	• Add organic matter (acidic, if possible) • For moderate alkalinity, add sulfur • For very high pH, determine cause through soil tests • Test for salt problems • Switch to low-salt fertilizers and amendments that acidify soil • Try foliar feeding for nutrient deficiencies • Don't use lime; use only gypsum if calcium is needed • Don't use cow manure if salt levels are high • Saline soils: Improve drainage and leach to reduce salt levels • Sodic and saline sodic soils: Improve drainage; treat with sulfur before leaching	chapters 4 and 5 pages 124, 152–153 pages 28–29, 152 pages 152, 154 chapter 6 page 115 pages 140, 157 page 154 pages 154–155 pages 152–157
Diseased soils	• Add organic matter (compost is best) • Practice good garden hygiene • Solarize soil • Use mulch to prevent soilborne plant diseases from splashing onto leaves • Choose resistant or tolerant varieties recommended for your area	chapters 4 and 5 page 158 pages 160–161 page 158; chapter 5 pages 158–159
Contaminated soils	• Add organic matter and keep soil mulched • Keep pH between 6.5 and 6.8 • Maintain abundant phosphorus • Get special soil tests for suspected contaminants • Determine source to limit further contamination	chapters 4 and 5 pages 150–153, 162 pages 118–119, 205 pages 162–163 pages 162–163

Improving Soggy or Poorly Drained Soils

Three things can cause poorly drained soils.

- **Topography** is the culprit if the surrounding land slopes toward the soggy spot: Water has no choice but to end up there after every rain.

- **Heavy or compacted soils** (which are often hard to distinguish) act like a plastic bag, holding water and not letting it drain away.

- **If the water table (groundwater) is close to the surface,** it can keep the soil above it saturated. Often these factors combine to keep soil soggy for long periods of time.

Raised Beds

One management strategy can help you grow great gardens regardless of the cause of your poor drainage. Raising the soil level at least 4 inches (10 cm) in raised beds often transforms a constantly wet site into an evenly moist site — the ideal condition for most types of gardens. Raised beds look nice, too, as plants are set off from their surroundings. And you can make them yourself.

Raised beds give roots room to grow without hitting the wet soil caused by a low spot or high water table. If the soil used to build the beds isn't too heavy, it will improve gardens in clay or compacted soils. For the best results with clay or compaction, though, double-dig the area before you build raised beds (see page 141).

Raised beds can benefit your garden in many other ways as well. They can improve soil drainage, soil aeration, and air circulation around plants; allow soils to warm up in spring; increase space available for roots in shallow or rocky soils; and in salty soils, create a low-salt zone for root growth.

MASTER GARDENING TIPS

A Few Perennials That Like Wet Feet

Many attractive perennials thrive with constantly moist or even wet feet.

Angelica *(Angelica archangelica)*
Astilbes (*Astilbe* spp.)
Blue flags *(Iris versicolor, I. virginiana)*
Candelabra primrose *(Primula japonica)*
Cardinal flower *(Lobelia cardinalis)*
Cattails (will overrun other plants) (*Typha* spp.)
Great blue lobelia *(Lobelia syphilitica)*
Japanese and Siberian irises *(Iris ensata, I. sibirica)*
Joe-Pye weed *(Eupatorium maculatum)*
Marsh marigold *(Caltha palustris)*
Many ferns (many spp.)
Swamp milkweed *(Asclepias incarnata)*
Yellow flag *(Iris pseudoacorus)*

Regrading

If you have enough money, you also have the option of regrading the topography of your yard or installing drainage pipes. Hiring a landscape professional is a good idea with either. While it's easy enough to fill small low spots with topsoil, regrading large areas could direct water into your basement or other unintended spot. Also, be aware that even small bulldozers tend to compact soil, especially if it's wet.

Drainage pipes usually work best if an experienced person lays them. They don't work if they're not sloped correctly. They tend to fill in over time, too, and have a habit of moving around from frost heaving. Before you hire anybody, try to find where the water is coming from. Sometimes a simple aboveground extension of a downspout, or redirecting the outlet from a downspout, will solve the problem.

Options for Raised Beds

For vegetable gardens, simply shape existing soil (if you double-dig it first, your soil will end up a few inches, or cm, higher to start). For large shrub or perennial borders, you may need to import topsoil or add a mix of half compost and half topsoil. If your climate is cool and/or humid, make beds 4 to 6 inches (10 to 15 cm) high for extra drainage. In warm or dry climates, aim for 3 to 4 inches (8 to 10 cm) to keep soil from drying out too much.

Vegetable gardens. Create temporary beds. Lay them out with stakes and string for tidy edges. Three to 4 feet is a good width; leave a path at least a foot wide between beds. Use a hoe to pull soil up into beds; rake the top level. If the lower soil in the paths tends to stay moist, mulch thickly or lay wide boards to walk on.

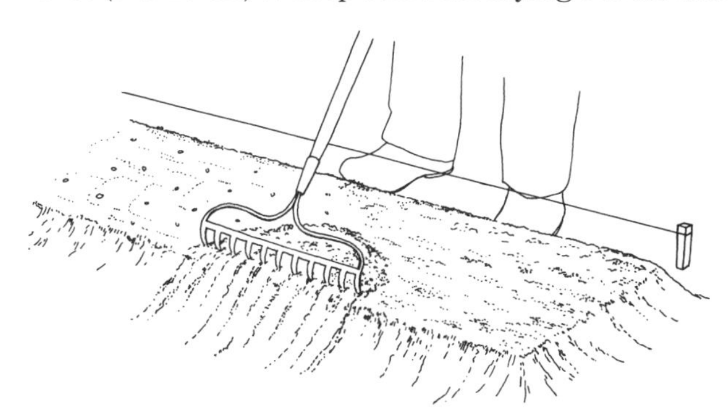

Permanent beds. Landscape timbers or 2 x 6 boards make good sides; more elaborate options include mortared brick and dry stone walls around raised beds. To keep landscape timbers from heaving in winter, drill holes through them at 5- to 6-foot intervals. Pound 2-foot lengths of slender steel pipe through the holes until ends are flush with the timber. Short pieces of pipe also make good stakes for holding wide boards against the raised soil sides. (Or use rebar, a concrete reinforcing rod, if you can get it cut to size.) You can buy raised-bed kits of plastic corner stakes with slots to hold wooden boards or recycled plastic "boards."

Trees and shrubs. Build up a low mound or berm (8 to 12 inches high) with gradually sloping edges before planting. For a single shrub, make it 3 to 4 feet in diameter; expand for beds of multiple shrubs. For a single tree, make it about as wide as the mature diameter for small trees, or half the mature width of large shade trees.

Improving Clay and Heavy Soils

Heavy clay soils can be made more loamlike by physical loosening with double-digging (see opposite page). Once loosened, you need to supply them with lots of organic matter to keep them that way. Since loam is a mix of sand, silt, and clay, you might be tempted to add sand to improve the texture of clay. Don't bother. Sand is very heavy, and you have to add several inches of the coarsest grade to make any difference. Spreading only a couple of inches of compost or other organic matter over soil gives better results than several inches of sand. Compost is a lot easier to deal with, too, as it's much lighter to cart and shovel.

The Benefits of Clay

Clay offers benefits to offset its heavy texture. It can hold large reserves of nutrients and moisture for plants. The main problem with clay soils isn't the clay itself but the high probability that such soils will become compacted, develop hardpan, or develop drainage problems. If you correct those problems, you can grow great gardens in clay. Since these problems are so often interrelated, you'll find that the double-digging described here can work wonders with compaction and poor drainage as well.

MASTER GARDENING TIPS

A Few Plants That Like Heavy Clay Soils

Perennials

Ajugas (*Ajuga* spp.)
Daylilies (*Hemerocalis* spp.)
Houttuynia *(Houttuynia cordata)*
Ornamental grasses (many spp.)
Pachysandra *(Pachysandra terminalis)*
Virginia creeper (vine) *(Parthenocissus quinquefolia)*

Shrubs

Barberries (*Berberis* spp.)
Creeping juniper *(Juniperus horizontalis)*
Euonymus (*Euonymus* spp.)
Viburnums (*Viburnum* spp.)
Willows (*Salix* spp.)
Rugosa rose *(Rosa rugosa)*
Shrub dogwoods *(Cornus alba, C. sericea, C. stolonifera)*
Spruces (*Pica* spp.)
Sumacs (*Rhus* spp.)

HINT FOR SUCCESS

Watch Your Magnesium

Soil chemistry can contribute to heavy texture and poor drainage. Too much magnesium acts like cement. If your soil tends to be heavy even though it's low in clay, test magnesium levels. They should be (and almost always are) much lower than calcium. If you get more than half as much magnesium as calcium, you could have problems.

On acidic soils with adequate magnesium, switch from using dolomitic limestone to calcitic limestone, which contains only calcium and no magnesium. Where magnesium levels are really high, you'll need to add gypsum (calcium sulfate) to correct the calcium-magnesium balance and loosen the soil. Gypsum has a reputation for loosening any heavy soils, but it really makes a dramatic difference only where unusually high magnesium levels compound the problem.

Double-Digging

To double-dig, you loosen and mix the first foot of soil, blending in amendments as you go, and then loosen the deeper soil in place. Do it a few weeks before you intend to plant, and its benefits will last from several years to indefinitely. It improves drainage and aeration, loosens compacted or heavy soils, improves root growth, encourages earthworms and other soil organisms, distributes nutrients and other amendments evenly through the root zone, and leads to healthier plants.

1 Outline the area with stakes and string. If the soil is very dry or very hard, water well and wait a day. Remove any existing sod and troublesome weeds. Spread a 2- to 3-inch layer of compost, leaf mold, or aged manure evenly over the area. Also spread lime or sulfur if needed to correct pH.

2 Start digging a trench about a foot wide and as deep as your shovel along one long side. Place removed soil in a cart or on a tarp so that it's easy to move.

3 Drive the tines of a garden fork or a broadfork as deep as you can into the bottom of the trench. Rock the handle back and forth to loosen the subsoil. Spread a little compost over the loosened soil.

4 Dig another trench alongside the first, turning the removed soil into the first trench and mixing in any amendments. Continue to the far edge of the bed. Fill the last trench with the soil from the first trench. Just before planting, rake the surface smooth.

Improving Compacted Soils and Hardpan

Compacted soils cause several problems. A thin compacted layer within looser soil, called hardpan, can have the same effect as soil that's completely compacted. In dry climates, natural rocklike hardpans called caliche are common. Any soil compaction often causes poor drainage. Roots can asphyxiate because there aren't enough pore spaces to allow air into soil; any existing spaces may be filled with water. Some soils and hardpan layers (especially caliche) are so dense that roots can't push through. These effects work together to restrict plant growth and health.

Some Solutions

Four techniques open up channels in dense soils with little or no danger of added compaction.

1. ***Earthworms.*** The first is the easiest: Encourage earthworms. If soils are only moderately compacted, earthworms may be able to loosen them for you.

2. ***Green manure plants.*** Use green manure plants with tough, deep roots. Growing a crop of alfalfa or sweet clover can help open up compacted soils.

3. ***Broadfork.*** Use a broadfork. It can loosen hardpan enough for earthworms to tunnel through, at least in spots.

4. ***Double-digging.*** In small areas, double-dig. It really helps to loosen soils and break up hardpans.

Avoiding Compaction

Any soil that's become compacted once is at risk of future compaction; clay soils are also at risk. Working these soils when wet can compress them into dense potter's clay, and they'll stay compressed once they dry. Wait until soil is moist, not soggy, to dig or cultivate. Moving heavy loads or even walking on such soils can cause compaction, so minimize both or wait until soil isn't soggy.

Minimize any activities that disrupt soil structure, especially tilling. To avoid creating a tillage pan or plow pan, don't turn over your soil unless you need to incorporate phosphorus, lime, or sulfur. Spread compost on the soil surface rather than tilling in leaves or other raw organic matter, or do each in alternate years. Try just raking rather than tilling to create beds for seeds or transplants.

Organic matter is a good shock absorber. Adding it improves aeration and drainage of compacted soils and helps keep soil loose after double-digging.

After years of adding organic matter, building up earthworm populations, and balancing your soil, you should see a major improvement in stability of soil structure, reduced compaction, and little or no hardpan.

HINTS FOR SUCCESS

Preventing Compacted Soils

- Never cultivate soil when it's wet.
- Don't step on beds when soil is wet (minimize even when dry).
- If you must walk on wet areas, lay down wide boards and walk on the boards to distribute your weight over a larger area.
- Install permanent paths on heavily traveled areas.
- Don't drive or haul heavy loads over wet soil or lawns.
- Disrupt soil structure as little as possible.
- Minimize use of rototillers to prevent formation of tillage pan.
- Choose lighter equipment such as push mowers rather than riding mowers or garden tractors.
- Maintain high humus levels, as spongy, porous humus resists compaction more than most soils.
- Encourage earthworms (abundant organic matter and minimal deep cultivation are the best ways to do this).

Using a Broadfork

Broadforks (also called U-bars) are great for breaking up shallow hardpan in lawns, in and around flower and shrub beds, and before planting shrubs or trees. Using it once or twice a season will allow more water to penetrate and increase aeration. Two types are available. The more common has a horizontal handle that you grip at both ends. The other type is shaped like a large U with vertical handles (hence the name U-bar). You can use a garden fork as a substitute, but it doesn't work as well and takes more effort.

Once again, make sure soil is only slightly moist before you start. Work gently around existing plants in established beds, and don't get too close to shallow-rooted trees and shrubs.

DID YOU KNOW?

Advantages of a Broadfork

- It can be used in small corners.
- You'll get faster results than from deep-rooted green manures.
- You can avoid the compaction caused by heavy equipment.
- It is much less expensive than hiring someone with a tractor and subsoiling attachment.

1 Grasp both handles and hold the tool vertically. Use one foot on the lower bar to push tines deep into the soil.

2 Pull back on the handle to rock the tines forward.

3 Pull tines out of the soil. Move a few feet away and repeat the process until you've covered the entire area.

HINT FOR SUCCESS

Planting Trees and Shrubs over Hardpan

Before you plant an expensive tree or shrub, dig a hole through the hardpan layer to the porous soil below, using a posthole digger or sharp spade. If the layer is really hard (such as caliche), you may need to rent a soil auger to bore through it.

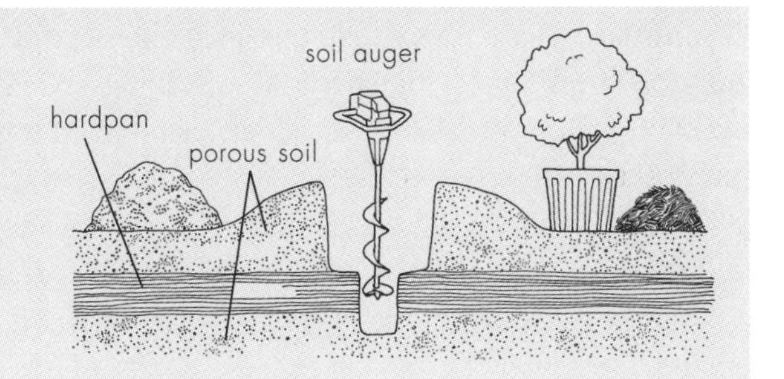

Improving Sandy or Excessively Drained Soils

While most plants prefer well-drained soil, it's possible to have too much of a good thing. Soil that drains too quickly is impossible to keep moist for very long, and applied fertilizers are quickly washed away.

Theoretically, adding a truckload of clay would improve the texture, but organic matter works just as well, is easier to handle, and is readily available. In addition to mixing organic matter into the soil, keep at least 2 inches on top at all times. Mulch slows down evaporation, keeps soil cooler, and contributes more organic matter.

Adding Nutrients

The nutrient problem requires an additional strategy. Instead of applying fertilizers all at once, feed plants smaller doses throughout the season. The sandier your soil, the smaller and more frequent the feedings. Try one-third the recommended amount (following label directions or amounts suggested in chapter 6) at the beginning of the season, an equal dose after a third of your growing season has passed, and the last dose two-thirds of the way through your growing season (in time for plants to use it before the season ends).

Watering

If you live where water supplies are limited, invest in a drip irrigation system. By delivering water efficiently, it minimizes nutrient leaching. Soaker hoses are an inexpensive alternative. Lay soaker hoses under mulch and leave in place for the season (disconnect from the hose leading to the water supply when not in use). That way you water the soil rather than the mulch.

Sunken Beds

If you have no problems with salt levels, creating a sunken bed is another good way to reduce watering. Surrounding sunken beds with low walls further increases their effectiveness by reducing air movement, which reduces evaporation. If you live in a windy area, you'll probably want to install a larger windbreak to shelter as much of your yard as possible from the drying effects of wind.

MASTER GARDENING TIPS

A Few Drought-Tolerant Plants for Sandy Soil

Perennials

Aloes (*Aloe* spp.)
Artemisias (*Artemisia* spp.)
Autumn sage (*Salvia greggii*)
Blanket flowers (*Gaillardia aristata, G.* x *grandiflora*)
Butterfly weed (*Asclepias tuberosa*)
Coneflowers (*Rudbeckia, Echinacea* spp.)
Euphorbias (*Euphorbia* spp.)
Evening primroses (*Oenothera* spp.)
Herbs
Mulleins (*Verbascum* spp.)
Ornamental grasses
Ornamental onions (*Allium* spp.)
Stonecrops (*Sedum* spp.)
Wildflowers, shrubs, and trees native to your region
Yarrows (*Achillea filipendulina, A. millefolium*)
Yucca (*Yucca* spp.)

Lawn grasses

Bermuda grass (*Cynodon dactylon*)
Blue grama grass (*Bouteloua gracilis*)
Buffalo grass (*Buchloe dactyloides*)
Zoysias (*Zoysia* spp.)

Building Sunken Beds

Sunken beds use topography to your advantage to collect water. They work best in hot, dry climates and reduce water use even for soils that don't suffer from excessive drainage. Build them only where there's no salt problem, though, as they can increase soil salt levels.

1 Divide the area into several smaller (3- to 4-foot-wide) pits for individual beds. (At high altitudes, create one large bed.) Remove the topsoil from each bed onto a tarp. If you can't tell the topsoil from the lower soil, remove at least the top 6 inches.

2 Excavate 18 to 24 inches (even deeper at high altitudes). Keep this material separate from your topsoil. If you encounter hard soil or caliche, use a heavy-duty mattock or pick, or hire the services of a backhoe. Mix the removed topsoil with an equal amount of organic matter and shovel it back in. Rake the surface smooth.

3 Shape some of the removed subsoil into low berms surrounding each bed. If it's as sandy as your topsoil, the best you can manage is a low sand dune. If it contains some clay, you can build it up a bit higher. If possible, build a low wall around entire area out of flat rocks, cinder blocks, or even hay bales.

DID YOU KNOW?

Benefits of a Sunken Bed

- Improves absorption of water.
- Improves water retention.
- Reduces evaporation from soil.
- Helps keep soil cooler in summer.
- Extends growing season at high altitudes.
- Simplifies irrigation.
- Reduces nutrient leaching.
- Reduces erosion.

Managing Slopes

Since water quickly runs down a steep slope instead of seeping in, soils on slopes tend to be drier than neighboring flat land. If your soil's heavy, slopes can be an asset, improving drainage. But if your soil is sandy, they just intensify the problem of keeping moisture and nutrients in the soil. Soil on slopes tends to be thinner and poorer than neighboring soil. Erosion leaches nutrients and soluble forms of humus. It also carries off exposed topsoil.

Tame with Terraces

Terraces are attractive landscape features that address all the above issues. In heavy soils, they give you level ground to garden in so you can take advantage of the improved drainage. If your soil's too well drained to begin with, the level areas slow down the flow of water so that it can seep in rather than just running off. To slow down the water even more, create retaining walls slightly higher than the soil level in the terraces. (See the following pages for directions for building terraces.)

Meadows Beat Mowing

If you don't mind the look of a grassy meadow, let the grass grow and reduce mowing to a couple of times a year. To win neighbors over to your meadow, plant lots of daffodils or other bulbs for a stunning spring display before the grass starts to grow. Long grass camouflages fading daffodil foliage.

MASTER GARDENING TIPS

A Few Low-Maintenance Perennial Groundcovers for Slopes

Sun-loving groundcovers

Bearberry *(Arctostaphylos uva-ursi)*
Cherry laurel *(Prunus laurocerasus)*
Cotoneasters (*Cotoneaster* spp.)
Creeping junipers *(Juniperus horizontalis)*
Creeping thyme *(Thymus serpyllum, T. praecox arcticus)*
Daylilies (*Hemerocallis* spp. and cultivars)
Forsythias (*Forsythia* spp.)
Heaths and heathers (*Calluna, Erica* spp.)
Houttuynia (*Houttuynia cordata* 'Chameleon')
Ice plant (*Delosperma* 'Alba')
Leadwort *(Ceratostigma plumbaginoides)*
Moss phlox *(Phlox subulata)*
Ornamental grasses (many spp.)
Pride of Madeira *(Echium fastuosum, E. candicans)*
Stonecrops (*Sedum* spp.)

Shade-tolerant groundcovers

Ajuga *(Ajuga reptans)**
Barren strawberry *(Waldsteinia ternata)*
Common periwinkle *(Vinca minor)*
Epimediums (*Epimedium* spp.)
Foamflower *(Tiarella cordifolia)*
Green-and-gold *(Chrysogonum virginianum)**
Hostas (choose spreading types) (*Hosta* spp.)
Lamium *(Lamium maculatum)**
Lily-of-the-valley *(Convallaria majalis)*
Pachysandra *(Pachysandra terminalis)*
Sweet woodruff *(Galium odoratum)*
Wild ginger *(Asarum canadense, A. europaeum)*

*also tolerates full sun

Lawn vs. Groundcovers

Lawn is a common option for steep slopes, but not a particularly good one, as mowing is so hard. If you don't want terraces, consider replacing lawn with low-maintenance groundcovers. The best groundcovers for slopes are vigorous plants that spread quickly and need little upkeep (see list on facing page). They should also be adapted to the soil, sun, and climate of your site.

Enrich the soil, at least in large planting holes, before setting out transplants. Spread burlap or a thick layer of straw over the area to hold soil in place and reduce watering (you can substitute other mulches, but choose ones that won't wash away). To use burlap (or erosion cloth, an open-weave version of burlap), spread it over the entire area. Cut Xs in the fabric to plant groundcovers. If the groundcovers haven't filled in completely before the burlap starts to break down, spread mulch between them until they fill in. Be prepared to supply extra water during the first growing season.

HINTS FOR SUCCESS

Strategies for Shallow Slopes

Any slope can erode, but a few tricks will keep your soil and nutrients in place.

- Cover any exposed soil with mulch.
- Run your rows along the slope, rather than up and down it.
- Minimize tilling, as loose soil washes away easily.
- Space vegetables and annuals close together so leaves are almost touching, to slow the speed of raindrops.
- If you don't want to build terraces, leave horizontal strips of lawn between beds to help hold soil in place. Or, grow nitrogen-fixing white Dutch clover as a permanent cover crop on horizontal paths; it tolerates mowing and foot traffic.

Terracing

Building low retaining walls (up to 1½ feet high) is relatively easy. Several shorter walls along a slope are much more stable (and much easier to build) than one or two tall walls. For higher walls, consult a landscape professional or builder. Large retaining walls must be well designed and anchored to keep them from washing out, cracking, or tumbling down and causing a landslide after a few years. It's worth the investment to make them well.

HINTS FOR SUCCESS

Advantages of Terracing

- Reduces soil erosion.
- Water seeps into soil instead of running off surface.
- Creates level beds that are much easier to work in.
- You can grow a much wider variety of plants than on slopes.
- Improves water retention.
- Minimizes nutrient leaching.
- Adds an attractive design element to yard or garden.

Temporary retaining walls. When planting groundcovers on a slope, build temporary retaining walls to hold the soil in place until plants get established. Drive 2-foot stakes partway into the ground to anchor the long, wide boards. Slender steel pipe makes good stakes; ask the hardware store to cut it for you.

HINT FOR SUCCESS

Mortared vs. Dry Stone Walls

Terraces made of dry stone walls fit well into most landscapes. Mortared stone or brick walls are more appropriate for formal landscapes. Mortared walls also require more expertise to build. For the best-looking and longest-lasting results, hire an experienced landscape professional or builder.

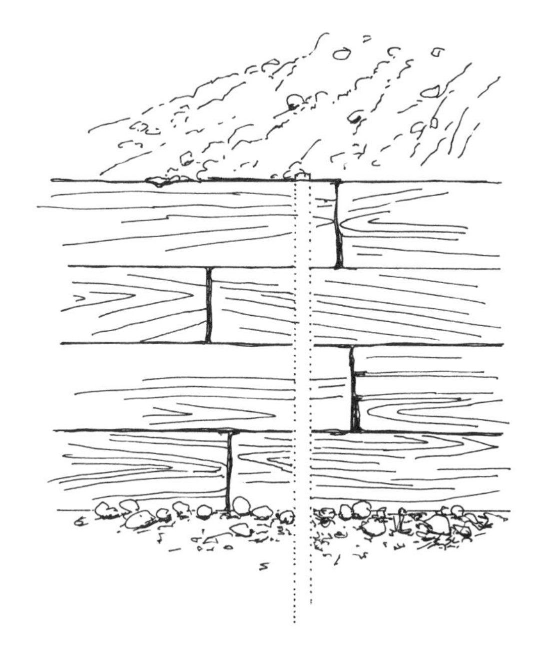

Permanent retaining walls. It's relatively easy to make permanent retaining walls out of landscape timbers. Anchor timbers in place and hold them together at the same time with steel pipe or concrete reinforcing bar 1½ ft. (46 cm) longer than the height of the wall. Drill a hole just wide enough for the anchor (to get a tight fit). Drive the anchor through all holes and into the ground until it's flush with the top timber.

Space the anchors about 6 feet (1.8 m) apart along the timbers. Leave a slight gap (weep hole) between the ends of the bottom timbers for good drainage.

Dry stone walls. Start at the bottom of the slope. Dig a trench at least 6 inches (15 cm) deep and fill with coarse gravel. Begin laying masonry, leaving small gaps (weep holes) every few feet along the bottom row to allow water to drain out.

Fill in behind the wall with a few inches (cm) of large gravel before adding any soil (especially important for heavy soils and where the ground freezes). Dig out soil from the slope to fill in and create a level bed.

Dig a second trench partway up the slope where you want your second wall. Repeat the remaining steps to finish.

Improving Acidic Soils

Acidic soils are common where rainfall is abundant. Most soils east of the Mississippi River and in the Pacific Northwest are acidic (though not those over limestone!). Slightly acidic soils (pH 6.3 to 6.8) are best for most plants and soil organisms because they ensure the greatest availability of essential nutrients. A simple pH test can tell you whether your soil is acidic, and to what degree.

Why Worry about Acidity?

Plants in very acid soils often show a combination of deficiency symptoms for nitrogen, phosphorus, potassium, calcium, magnesium, and/or molybdenum because these nutrients are locked up. Organic matter takes longer to break down in acidic conditions because the activity of soil organisms slows. As a result, the nutrient supply is further reduced.

In very acidic soils (pH below 6), aluminum can become so abundant that it harms or kills plants. Even iron and manganese can reach harmful levels at low pH. Extractable aluminum is measured by professional soil tests to provide more information about soil acidity. If your soil's naturally low in aluminum, your plants won't suffer as much if the pH drops.

Adding Lime

Periodic liming of acidic soils can replace the neutralizing calcium and magnesium that get washed away by rain. Don't add lime unless a soil test shows that you need it, though. Overliming causes problems that are worse than acidity.

Fall is the best time to spread limestone, because it takes a while to become available in the soil. The more finely ground, the faster it works; make sure the label says that at least 50 percent of it will pass through a 100-mesh screen. Pelletized limestone, which is less dusty, is larger but designed to break down quickly. To avoid shocking soil organisms, add no more than 5 pounds of limestone per 100 square feet at one time. Wait six months (and ideally retest pH) before adding more.

More Solutions

A couple of other tricks will also help you manage soils that are too acidic. Maintaining abundant amounts of organic matter increases both the buffering capacity of soils and plants' ability to tolerate some acidity. Avoid acidic forms of organic matter (pine needles and peat moss). Well-aged manure is good because it's slightly alkaline and supplies both calcium and magnesium. Also avoid acidifying fertilizers such as triple superphosphate. Switch to nutrient sources that are slightly alkaline (bonemeal, rock dusts, and guano) or neutral (Sul-Po-Mag).

MASTER GARDENING TIPS

A Few Plants That Like Acidic Soil (pH 5–6)

If your soil is really acidic and you don't want to spread lime over your entire garden every few years, choose plants that are naturally adapted to acidic soils. There are many more plants for acidic soils; your Cooperative Extension Service can recommend varieties and species particularly suited to your area.

Food crops

Apples
Blueberries
Cashews
Cranberries
Gooseberries
Potatoes
Raspberries
Strawberries

Perennials

Baptisia *(Baptisia australis)*
Chrysanthemums (*Dendranthema morifolium* and cultivars)
Cimicifugas (*Cimicifuga* spp.)
Lilies (*Lilium* spp. and cultivars)
Lily-of-the-valley *(Convallaria majalis)*
Trilliums (*Trillium* spp.)
Turtleheads (*Chelone* spp.)

1 For lawns and large areas, use a handheld or push-type fertilizer spreader. For small beds you can broadcast by hand. Wear gloves and a dust mask.

2 Work materials thoroughly into the top 6 inches of soil with a garden fork or small tiller.

3 Retest soil after a year and apply more lime if needed. Once you reach the 6.3 to 6.8 range, you'll only need to test every four years or so.

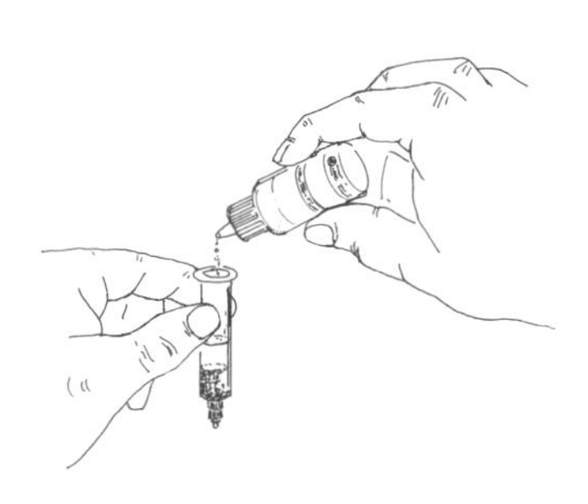

Master Gardening Tip

How Much Lime to Use

The following recommendations are for approximate amounts of lime needed on 100 square feet of soil to raise pH from 5.5 to 6.5. Start with these amounts, and retest your soil before adding any more.

Sandy soils need less calcium to correct acidity. You need to spread lime on them more often, though. Clay soils have large reserves to fill and therefore need more calcium, but you can wait longer before adding more. Soils rich in organic matter also need more lime; double the recommended amount for such soil. If your soil's very low in organic matter, reduce amounts by about 25 percent until you've built up your soil's humus.

Calcitic limestone supplies only calcium; it's the best choice for most soils. You can substitute finely ground oyster or clam shells for calcitic limestone. Use dolomite limestone only if your soil tests low in magnesium. Too much magnesium locks up other nutrients.

Wood ashes, which are rich in potassium, will raise soil pH, but don't add fertilizer containing potassium (such as 5-10-10) or you'll throw off your soil balance. Test calcium, magnesium, and potassium levels periodically. Fast-acting wood ashes can burn plants and soil organisms; add them at least three weeks before planting, and mix in well. *(For metric equivalents, see "Useful Conversions" on page 208.)*

Approximate Amount of Lime Needed per 100 Square Feet to Raise pH from 5.5 to 6.5

Soil Texture	Calcitic Limestone	Dolomite Limestone	Wood Ashes
Light (sandy)	2½ lbs.	2–3 lbs.	3–4 lbs.
Medium (loamy)	6½ lbs.	6 lbs.	8 lbs.
Heavy (silt or clay)	9 lbs.	7–8 lbs.	9–10½ lbs.

Improving Alkaline Soils

Alkaline soils are common in drier climates and therefore in western states. But just as pockets of alkaline soil occur in the East, pockets of acidic soil occur in the West where sulfates or other acidic compounds are abundant. You can't always tell by geography or rainfall.

High Salt or Sodium Levels

There's an even more important reason for testing your soil in alkaline-prone areas: to find out whether your soil carries high salt or high sodium levels. Either complicates soil management. If a soil test shows high salt levels or if white, salty crusts ever appear on the surface of your soil, skip ahead to the section on saline soils (page 154). If your pH is over 8.5, you've probably got high sodium levels that make it hard to grow almost anything; skip ahead to the discussion of sodic soils (page 156). If you're not sure, read both.

Adding Sulfur and Organic Matter

Moderately alkaline soils are managed much like acidic ones, except sulfur is substituted for limestone to lower the pH. Powdered sulfur (also called elemental or agricultural sulfur) is the best form to use. It works in six to eight weeks, lasts six months or more, and it's pure sulfur. Aluminum sulfate is faster-acting but doesn't last as long and can harm soil organisms. Iron sulfate is similar but supplies iron.

Again, organic matter increases both the buffering capacity of soils and the tolerance of plants. Seek out acidifying types, such as oak leaves, leaf mold, ground bark, aged sawdust, peat moss, pine needles, and pine or cypress bark mulch. While they can take three months to a year to work, their benefits last much longer than do any of the forms of sulfur.

MASTER GARDENING TIPS

A Few Plants That Like Slightly Alkaline Soil (pH 7–8)

Food crops

Asparagus
Avocado
Beet
Cabbage family
Green beans
Lettuce
Melons
Okra
Onions and leeks
Peach
Spinach
Swiss chard
Walnut

Perennials

Avens (*Geum* spp.)
Baby's-breath (*Gypsophilia paniculata*)
Coral bells (*Heuchera* spp.)
Peonies (*Paeonia* spp.)
Pinks (*Dianthus* spp.)
Rock roses (*Helianthemum* spp.)

Choose Fertilizers Carefully

Many alkaline soils have larger nutrient reserves than acidic soils; they may not need all the nutrients in a typical balanced fertilizer. Avoid fertilizers that increase alkalinity; switch to acidifying types such as cottonseed meal, a slow-release form of nitrogen. Concentrated, synthetic fertilizers may be helpful if used in small amounts. Ammonium sulfate supplies nitrogen and lowers pH about 75 percent as much as sulfur. It can burn plant roots and soil organisms, though, and its effects last only weeks in very alkaline soils.

Alkaline soils lock up many micronutrients out of plants' reach. Foliar feeding is the best and fastest way to correct a micronutrient deficiency. Liquid seaweed (or a combination of seaweed and fish emulsion) is a good choice.

More Solutions

Where soils are extremely alkaline, save your frequent neutralizing efforts for the plants that need it most, like tomatoes and peppers. Or sink large containers in the ground and fill with a mix of compost and imported, neutral topsoil. In the long run, importing topsoil may be cheaper than acidifying amendments.

A Simple pH Test

You can quickly check whether your soil is alkaline or highly acidic using just vinegar and baking soda. Get a professional soil test (see chapter 2) for a more accurate pH reading before attempting to correct either type of soil.

1 Dig up a handful of soil. Remove any pebbles and large bits of plant debris. Let soil air-dry. Crush any large crumbs with the back of a spoon.

2 With a large stainless-steel or plastic spoon, scoop up a tablespoon of the dry soil. Add several drops of vinegar. If this makes your soil fizz, it's alkaline. The pH is over 7.5.

3 Scoop up another tablespoon of soil and add water until it's very moist. Add a pinch of baking soda. If it fizzes this time, it means your soil has a pH of less than 5, which is too acidic for most plants.

Master Gardening Tips

Using Sulfur to Lower pH

The following recommendations are for approximate amounts of sulfur needed on 100 square feet of soil to lower pH from 7.5 to 6.5. To reduce stress to both plants and soil organisms, add no more than 1½ pounds of sulfur per 100 square feet at one time (for heavy soil, 2 pounds). If you need to lower pH by more than one unit, spread 1 pound in spring and another in fall. Retest soil before spreading any more the following spring. Test pH of alkaline soils annually.

Spread sulfur in the same way you would lime (see page 150). *(For metric equivalents, see "Useful Conversions" on page 208.)*

Approximate Amount of Sulfur Needed per 100 Square Feet to Lower pH from 7.5 to 6.5

Soil Texture	Powdered Sulfur	Aluminum Sulfate	Iron Sulfate
Light (sandy)	1 lb.	2½ lbs.	3 lb.
Medium (loamy)	1½ lb.	3 lbs.	5–5½ lbs.
Heavy (silty or clayey)	2 lbs.	5–6 lbs.	7½ lbs.

Improving Saline Soils

Soils with high salt levels, referred to as saline, may be caused by naturally occurring mineral deposits in dry climates. They can also be created by irrigation where dry air causes high evaporation rates. Repeated use of high-salt fertilizers and/or dehydrated manure from feed-lot cows (which are fed lots of salt) can cause problems in dry climates or on susceptible soils. Poor drainage is almost always a contributing factor.

Total salt levels are measured by how well soils conduct electricity. Pure water doesn't conduct well; the more salt it contains, the better it conducts. An electrical conductivity (EC) value above 4 on a professional soil test indicates your soil needs leaching. Values of 2 to 4 start to interfere with plant growth.

HINT FOR SUCCESS

Check Your Water

If your soil is salty, check whether your water source is salty, too. You could be increasing soil salinity every time you water. Don't use softened water for leaching or watering; it's too high in sodium. Install a faucet that bypasses your water softener or set a barrel under a downspout to collect rainwater.

Correct by Leaching Soil

The idea behind leaching is to flush the salts out of the soil. You need to apply a lot of water — 6 inches of water will leach half of the salt out of the top foot of soil. Leaching can't work unless your soil's well drained, so break up hardpan or caliche and improve drainage before you start (see page 138). Also, make sure you don't have high sodium levels (if sodium is high, see page 156). By getting your soil tested for sodium, you'll also get professional advice on leaching.

MASTER GARDENING TIPS

A Few Plants That Tolerate Salty Soils

Food crops	*Green manures*	*Flowers*	*Woody plants*
Asparagus	Barley	Asters (*Aster* spp.)	Date palm *(Phoenix dactylifera)*
Beet	Birdsfoot trefoil	Blanket flowers (*Gaillardia* spp.)	Elms (*Ulmus* spp.)
Broccoli	Oats	Coreopsis (*Coreopsis* spp.)	Fig (*Ficus carica)*
Cabbage	Rye	Phlox (*Phlox* spp.)	Grapes (*Vitis* spp.)
Cantaloupe	Sorghum	Portulaca *(Portulaca grandiflora)*	Honeysuckles (*Lonicera* spp.)
Cauliflower	Sweet clovers	Sedums (*Sedum* spp.)	Hydrangeas (*Hydrangea* spp.)
Kale	Wheat	Statices (*Limonium* spp.)	Junipers (*Juniperus* spp.)
Rosemary		Yuccas (*Yucca* spp.)	Oaks (*Quercus* spp.)
Spinach	*Lawn grasses*	Thrifts (*Armeria* spp.)	Olive *(Olea europaea)*
Squash	Bermuda grass		Privets (*Ligustrum* spp.)
Watermelon	Perennial ryegrass		Willows (*Salix* spp.)
Zucchini	St. Augustine grass		
	Tall fescue		

Leaching to Remove Salts

Try leaching at the beginning or end of the growing season. That way you won't be washing fertilizer away. Do it as early as possible in the day to minimize evaporation of the water you're applying. Let soil dry out slightly before cultivating or planting.

1 Set up an ordinary sprinkler in the center of the area to be treated. Place a rain gauge or bucket halfway between the sprinkler and the edge of the area. Adjust either until it's level.

2 Turn on the sprinkler, adjusting the water pressure until the sprinkler sprays water over entire area. Check to make sure water is falling into your gauge.

3 If you see puddles or any water flowing from the site, turn off the sprinkler until the water soaks in. Continue until the water in the gauge reaches 6 inches (15 cm). Expect this process to take a while; most soils can absorb only 1 to 2 inches (2.5 to 5 cm) per hour.

HINT FOR SUCCESS

Going Native

In any climate, growing plants that are native and therefore adapted to the area reduces work. In the difficult soils of the western deserts, this is especially true.

For information on native plants, contact either of the two agencies listed below:

National Wildflower Research Center
4801 LaCrosse Avenue
Austin, TX 78739
512-292-4200
Web site: www.wildflower.org

Native Seeds/SEARCH
2509 N. Campbell Avenue, Box 325
Tucson, AZ 85719
520-622-5561

Managing Sodic Soils

Soils with high sodium levels, called sodic soils, are characterized by extreme alkalinity (pH of 8.5 and up). On top of that, the high sodium levels are toxic to many plants and can completely destroy soil structure. Sodium works on soil like detergent on grease: It completely disperses the particles. It breaks up any crumbs and plugs up drainage, leading to compacted soils that repel water. Sometimes only a small area or "slick spot" is affected, surrounded by more normal soil.

The treatment described on page 157 can convert sodic soils into manageable soils. Your soil will still be alkaline, though, and may retain enough sodium to cause problems for some plants. Choose plants that can handle some alkalinity, and grow salt-sensitive plants in containers.

Combination saline-sodic soils are easier to work with because they behave more like saline soils. They contain enough calcium and magnesium to counteract the sodium. But ordinary leaching of saline soils high in sodium is a bad idea: You can wash away calcium and magnesium and instantly convert them to difficult sodic soils.

MASTER GARDENING TIPS

Salt-Sensitive Plants

If you have saline or sodic soils, avoid growing any of the following plants directly in the ground.

Food crops
Carrot
Celery
Corn
Cucumber
Green, lima, and kidney beans
Onion
Potato
Rhubarb
Raspberry
Most fruit trees

Green manures
Alsike clover
Crimson clover
Red clover
White Dutch clover

Lawn grass
Kentucky bluegrass

If you've got sodic soils, you've probably already discovered one of the easiest solutions: growing plants in containers of imported topsoil or soilless potting mix.

Using Gypsum

A more involved form of leaching is needed to remove sodium. Treating the soil first with gypsum replaces sodium in soil reserves with calcium, freeing up the sodium so it's easily washed out. About 5 pounds of gypsum is needed for 100 square feet on loamy soil; if your soil's heavy, use 10 pounds.

1 Spread gypsum evenly over the area to be treated. Mix it thoroughly into the top few inches of soil to eliminate any untreated pockets. This may take some effort in hardened sodic soils.

2 Thoroughly moisten the mixed soil. Moisture is needed for the chemical reaction converting sodium into a readily dissolved (and leached) form. In dry climates, water in the morning or evening so that less water evaporates. Allow the soil to sit for a day or two to let the chemical reaction proceed. Water again, if needed, to keep soil moist. Then proceed with leaching, as described on page 155.

MASTER GARDENING TIP

Saline or Sodic? Know Your SAR

SAR stands for sodium absorption ratio and tells you how much sodium is present in comparison to combined calcium and magnesium. The latter two help protect soil from the structure-destroying effects of sodium.

A professional soil test will give you an SAR value. If your SAR is below 4, you don't have to worry. Levels of 4 to 6 start interfering with plant growth. Values between 4 and 13 mean a soil is saline. Values of 13 and above mean a soil is either sodic or a combination of saline and sodic. If the SAR is over 13 and the pH is over 8.5, it's sodic. If the pH is 8.5 or lower, the soil is classified as saline-sodic (which is easier to work with than sodic).

Controlling Soil Diseases

Several soilborne diseases trouble plants. These include wilts that cause plants to keel over even in moist soil, and several leaf-spot diseases. They infect plants either directly through the soil or by splashing up onto leaves. The wilts have no cure and often kill plants. Some leaf-spot diseases can be kept in check with compost tea or fungicides, but they can't be cured. Prevention is the best control.

Many of these diseases are troublesome only in certain regions. Oak root fungus *(Armillaria)* is a problem mostly in California; it exists elsewhere but doesn't seem to cause the same damage. Contact your local Cooperative Extension Service for help identifying any diseases. People there can also supply names of resistant plant species or varieties that grow well in your area.

Use a Three-Pronged Attack

The combination of crop rotation, maintaining high levels of organic matter, and mulching (plus solarization for severe problems) allows you to manage or control all sick soils. Using this multiple approach provides the best, longest-lasting control. In contrast, strong chemicals kill off beneficial as well as harmful soil organisms; as a result, they provide only short-term control and require frequent use.

- **Rotating crops in vegetable gardens** effectively starves nematodes and diseases; it keeps large populations from building up.
- **High levels of organic matter** shift the balance away from troublesome organisms in favor of the helpful ones, some of which feed on the troublemakers. Experiments have shown dramatic reductions of nematodes in response to incorporating any organic matter; adding eggshells, seafood meal, or ground shells of shellfish gives even more dramatic results.
- **Mulching helps reduce wilting** and heat stress for plants that are already diseased. It can prevent some diseases. Mulch keeps leaf-spot fungi in soil from splashing onto leaves of healthy plants.

Solarization and Other Solutions

For gardens with severe disease or nematode problems, solarization (using the sun's heat to kill diseases) can provide control for up to three years. As a bonus, it kills many weed seeds. Good garden sanitation keeps you from spreading nematodes or diseases to uncontaminated soil. Finally, choosing plants that are unaffected by the disease, or resistant varieties of susceptible plants, guarantees you good gardens no matter what.

MASTER GARDENING TIPS

How to Treat "Sick" Soils

- Rotate annual crops.
- Maintain high levels of organic matter (compost is best).
- Keep soil mulched (except in cases of root rot).
- Improve soil drainage to reduce root rot.
- Solarize soil.
- Grow resistant varieties, or plants that aren't susceptible.
- Grow very susceptible plants or varieties in containers on a patio or deck or other location where contaminated soil can't splash onto them.
- Practice good garden sanitation: Remove and destroy diseased plants, and clean tools after digging in sick soil.

HINT FOR SUCCESS

Disease-Resistant Tomatoes

Many varieties of tomatoes have been bred for multiple disease resistance. The diseases are identified by capital letters after the variety name, as in 'Celebrity' hybrid VFFNT, which resists verticillium wilt, two strains of fusarium wilt, nematodes, and tobacco mosaic virus.

Here's a list of what the letters mean:

V: Verticillium wilt
F: Fusarium wilt
FF: Fusarium wilt, races 1 and 2
N: Nematodes
T: Tobacco mosaic virus
A: Alternaria blight (early blight, a leaf-spot disease)
S: Stemphylium (gray leaf spot)

Nematodes (eelworms) are a pest rather than a disease, though they cause symptoms resembling mild cases of wilt diseases. Most of these wormlike organisms are very small to microscopic. While some are beneficial soil organisms, others, such as root-knot nematodes, harm or kill plants. If plants are stunted and wilt even though the soil is moist, pull up the roots to look for small, beadlike bumps.

Solarizing Soil

Solarization is the process of using the sun's heat to kill off many diseases and weed seeds. Some of the soilborne plant diseases controlled by solarization include fusarium wilt, verticillium wilt, early blight (alternaria), root rot, damping off, crown gall, and nematodes. Fortunately, beneficial soil organisms can take hotter temperatures than can harmful ones, so this process doesn't destroy the good guys. (Or, at least, enough survive to quickly repopulate the area.)

For the most effective disease control, the soil temperature should reach 114°F (46°C) for three to eight weeks. Lower temperatures will still kill many weed seeds and may also slow the diseases long enough to give your plants the head start they need to get through the growing season.

This technique greatly reduces weeds for at least a year. Minimize tilling or disturbing the soil after solarizing. That way you'll avoid stirring up healthy weed seeds (or nematodes) from deeper in the soil.

Where and When to Solarize

Solarization works best with bright sunlight and warm temperatures. It's most effective in Arizona, southern California, and other regions with hot, sunny, dry summers. Where clouds and cooler temperatures are common, use a second layer of plastic to generate more heat. Choose the warmest, sunniest time of year. In many places this is July and August. Where summer thunderstorms are common, solarizing somewhat earlier or later may give you a longer period of available sunlight. *(For metric equivalents, see "Useful Conversions" on page 208.)*

HINT FOR SUCCESS

Hot Tip for Cool Climates

Using chicken manure along with solarization can generate additional heat so that only one layer of plastic is needed. In one experiment, 37 pounds of fresh chicken manure were used for 100 square feet of garden. The chicken manure increased the soil temperature 10 to 25°C at a depth of 6 inches. The high temperatures gave 100 percent control of troublesome weed seeds in the top 6 inches. In similar studies, the chicken manure also helped control nematodes.

You can use any fresh (or only slightly rotted) manure if you don't have access to chicken manure. Spread 2 inches deep before turning, and mix it in well.

MATERIALS

- Clear plastic sheeting (2 to 4 mil thick)
- Shovel or tiller
- Rake
- Soaker hose
- Soil stakes or heavy stones
- Empty soda cans (for cooler climates)

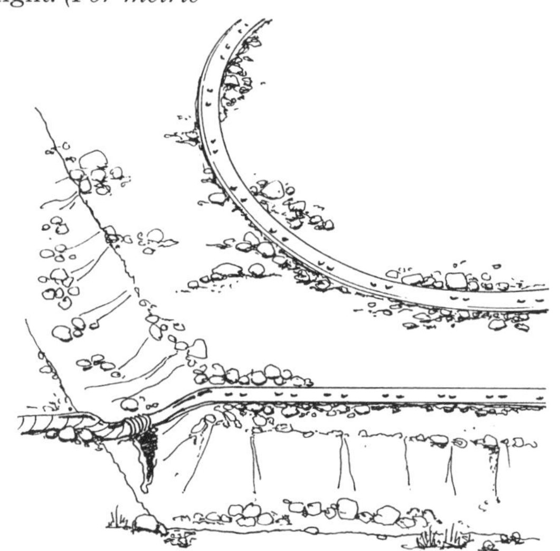

1 Dig up or pull any plants. Rototill or turn over the soil to a depth of 12 inches. Remove stones, sticks, and other large debris. Rake the surface until smooth, so you won't have to disturb the soil after it's been treated.

Water the soil until the top 2 feet are well moistened but not soggy. If your soil is very sandy and well drained, it'll need additional moisture during the procedure, so lay a soaker hose on top of the soil at this point.

2 Spread a sheet of clear plastic over the entire area. Leave it somewhat loose to allow room for it to puff up with steam. On raised beds, make sure the plastic drapes over and covers the edges. On flat or sunken beds, dig a trench around the area and bury the edges in the trench. Use soil stakes or heavy stones to hold the plastic in place. In the sunny Southwest, this layer of plastic is all you need.

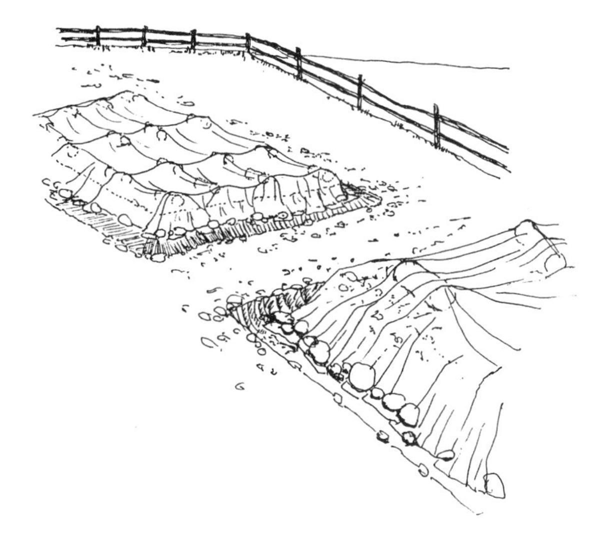

3 In climates that are cloudy, cool, or humid, spread a second layer of plastic. Evenly space empty aluminum cans every 4 to 6 feet across the first layer. Spread a second layer of plastic on top. Rearrange the cans if necessary; the second layer should float on top of the first layer (except at edges). Secure the edges of the second sheet under soil and heavy stones or stakes. (For small beds, you can substitute bubble wrap for the bottom layer of plastic and skip the cans, but the soil won't heat up quite as much.)

4 Let the soil bake. In the Southwest, wait three to four weeks to remove the plastic. In other areas, wait at least six weeks and preferably nine. If your soil is very well drained, lift the plastic after a couple of weeks to see if the soil is still moist underneath. If not, run the soaker hose to provide moisture.

Contaminated Soils

Contamination is more serious than other soil problems because it poses a health hazard for humans, not just for plants. High levels of some contaminants can be toxic to plants and soil organisms. Levels that are low enough to go undetected in plants can still cause harmful levels to accumulate in humans who eat the produce. If you're growing only ornamentals, not edibles, there's still a potential health hazard. High levels of some contaminants such as lead can be harmful to gardeners or to children playing on the soil, especially when the soil's dry and dusty.

Have the Soil Tested

The first step if you suspect contamination is to get your soil tested. Massachusetts and Rhode Island, as well as several private soil labs, check lead levels as part of their basic soil test. In other states, you'll need to request and probably pay extra for special tests. There isn't a single, simple test to tell whether any contaminants are present; each requires its own test.

To avoid spending lots of money on unnecessary tests, do a little research first. Find out what industrial or agricultural activities went on in your area or on your property to narrow the list of suspects. Identifying the source of contamination is also important if you want to stop any further or future problems. If your soil turns out to be contaminated, you'll need to retest it periodically to make sure your management techniques are working.

Managing Moderate Cases

Moderate cases of contamination are fairly easy to manage. Rather than removing soil, change the soil environment to reduce the availability of the chemical(s). Three strategies will lock up most chemicals out of plants' reach.

- **Organic matter** bonds tightly with heavy metals and other contaminants, so maintaining high levels helps keep the bad guys locked up.

- **Phosphorus** is also good for binding up most metals to keep them out of reach, so keep the soil well supplied with phosphorus. (For one contaminant — arsenic — phosphorus isn't a good idea. Arsenic isn't a metal, and more phosphorus encourages plants to take up more arsenic. Where it's a problem, add small amounts of zinc and iron instead, but be careful, because too much of either causes other problems.)

- **Maintain a pH of 6.5 to 6.8.** Acidic soils increase the availability of most contaminants, but if you make soil too alkaline, essential nutrients become unavailable to plants.

MASTER GARDENING TIPS

Strategies for Treating Contaminated Soils

- Keep pH at 6.5 to 6.8.
- Maintain abundant organic matter.
- Maintain somewhat high phosphorus levels (except where arsenic is present).
- Keep soil mulched to minimize dust on plants.
- Where contamination is minimal, peel root vegetables and remove outer leaves of leafy vegetables before using; rinse all vegetables well.
- Where contamination is severe, avoid growing root or leafy vegetables (or grow them in containers of soilless potting mix).
- Where lead levels are high, grow broccoli, cauliflower, kale, and/or collards, as these absorb very little lead from the soil. (Be sure to wash well, though.)

How to Minimize Risks

A few simple techniques will minimize health risks. Keep the soil mulched to keep dust where it belongs — on the ground. Steer children away from contaminated areas; it's hard to keep play areas covered with mulch and hard to keep children from stirring up dust as they play. Peel root vegetables, remove outer leaves of leafy or heading vegetables, and wash all produce well to minimize your consumption of contaminants. Test your soil regularly and amend as needed; if the pH slips to more acidic levels or if organic matter levels drop, contaminants can escape their lockup relatively quickly.

Handling Severe Cases

For severe cases of contamination, enlist the help of your local Cooperative Extension Service. People there may refer you to your state environmental agency in some situations. Your health is at risk, so it's worth seeking experienced help. Where soils contain high lead levels, the top few inches (cm) of soil may require removal by experienced professionals to avoid simply spreading the contamination. Don't grow any food crops until contamination is reduced to safe levels. The Cooperative Extension Service can tell you what levels of different contaminants are safe for food crops. They'll also tell you which ornamental plants can tolerate the contamination levels present in your soil.

DID YOU KNOW?

What's a Heavy Metal?

Heavy metal is a term used loosely to distinguish potentially harmful metallic elements from beneficial ones. Iron, magnesium, copper, zinc, molybdenum, cobalt, manganese, potassium, and calcium are all classified as metallic elements, and as essential nutrients. All are beneficial unless present in unusual, extremely high levels.

Metals are called heavy if they're denser and therefore heavier than metals such as magnesium and aluminum. The term is often used only for those heavy metals considered contaminants, especially lead, mercury, cadmium, silver, and chromium. All are toxic to animals and plants in relatively small doses. The essential nutrients copper, zinc, and nickel can also be considered heavy metals, because they're both dense and, at high levels, potentially toxic.

MASTER GARDENING TIPS

When to Test Your Soil for Contaminants

- If you want to grow vegetables next to an old house or other building, especially if you know it used to be painted white, test for lead. (Until fairly recently, lead was the pigment in most white paint.)
- If you want to grow vegetables near a road, test for lead. (Exhaust from leaded gas may have increased amounts in soil nearby.)
- On old orchards, old cotton fields, and old tobacco fields, test for arsenic, lead, and copper. (Before the 1940s, these compounds were used in the most common pesticides.)
- If the site was used for mining or smelting, test for arsenic and heavy metals such as nickel, cadmium, copper, or mercury. (Try to research which metals were mined or used in processing.)
- If you know your property was formerly the site of an incinerator or industrial plant, test for specific chemicals. (First you'll have to research what chemicals were used, as each requires a different test.)
- If sewage sludge was spread repeatedly on the site, test for cadmium and excess zinc. Concentrations of metals in composted sludges have been reduced in the past 20 years, so older applications are most apt to be problematic.

Thin or Stony Soils

Some sites are so rocky that they're covered with little, if any, soil. In areas of the West and Southwest, the baked soil is so hard that shovels bounce off. Other areas that were once covered with topsoil have been stripped by bulldozers during construction or by severe erosion.

If your subsoil is soft enough to dig, you can use the soil-improving techniques described throughout this book. A few successive crops of green manures can transform subsoil into acceptable topsoil. While deep-rooted types are best, ryegrass is faster. You can squeeze several crops of rye into a single season.

Extend Your Soil

Figure out where the soil is deepest and start gardening there if it's within reach of water. Make as much compost as you can. Use liberal amounts of compost to extend what little soil you have. Mixing homemade compost with imported soil will stretch it and save you money. As you build up measurable soil, or as your subsoil is undergoing transformation, watch for other problems such as poor or excessive drainage and acidic, alkaline, or salty conditions. Read the appropriate sections of this chapter to make the most of what you've got. Above all, watch out for erosion (from wind or water); mulch or terrace as needed to protect your precious resource.

MASTER GARDENING TIPS

Strategies for Thin or Stony Soils

- Build raised beds using imported topsoil and lots of compost.
- Grow plants in containers.
- Design a garden featuring rock-garden plants.
- Landscape with native plants.
- Contact local public gardens or your Cooperative Extension Service for information on plants and techniques suited to your region and your specific site.
- Look for pockets between boulders that you can fill with soil to support a few plants.
- Hire a jackhammer to hollow out a few planting holes.
- Make as much compost as possible so you can slowly increase the soil depth in various areas.

In many cases, you'll have to import soil in order to grow anything beside native plants (or those from similar regions that are well adapted to your conditions). Grow in containers until you've built up pockets or beds of soil deep enough for gardening. If rainfall is scant, rig up a drip system to each container so you don't spend all your gardening time and energy watering.

CHAPTER·8

Fine-Tuning Tips for Specific Plants

In all types of gardens, plants have different soil needs. Some prefer acidity, others alkalinity. Plants that require a rich diet, called heavy feeders, fare poorly with a lean diet or in poor soil. Those that like a lean diet (light feeders) mope with added fertilizer or in very rich soil. But the better your overall tilth, the more flexible your plants will be about their requirements, from pH to fertilizer.

Whether they're heavy or average feeders, plants grown for their leaves — ornamental foliage plants and leafy vegetables — need more nitrogen. (Corn does, too.) If you're choosing a balanced fertilizer, feed them with one whose N-P-K analysis is roughly equal (5-4-3, 5-5-5). Or, if you're using compost, add some manure to boost nitrogen. Plants grown for their flowers, pods, or fruits need less nitrogen and more phosphorus. They prefer formulas with lower N and higher P numbers (2-4-2, 5-10-5). Root vegetables also thrive with less nitrogen but a little extra potash (5-10-10 or 6-8-8, for example).

Balanced Fertilizers

If your soil has a good nutrient (and pH) balance and is enriched with organic matter, you'll probably get good results from all types of plants with any all-purpose balanced fertilizer. Three commonly recommended synthetic formulas are 5-10-5, 5-10-10, and 6-8-8. Most balanced organic fertilizers have lower formulas but make up for this by supplying micronutrients. Any blend with close to equal numbers and preferably slightly less nitrogen can be considered all-purpose. The application rates on fertilizer labels are for mediocre soils. Once you've achieved moderately fertile soil, try reducing recommended rates by one-third. Be guided by how well your plants perform.

In This Chapter

Garden Problem-Solver

Don't just assume your plants need fertilizer if they're not doing well. This chart describes common garden problems that may be caused by soil conditions. There isn't room here for all possible problems, or all possible treatments, but this chart will point you in the right direction.

It's hard to generalize plant needs throughout the country. Use the pH levels and other information here only as a guide. Your local Cooperative Extension Service and state agricultural university can provide you with information tailored to your soil and climate.

Symptom	Possible Causes	Treatment/ Prevention
Young plants die	Fertilizer burn	Switch to slow-release fertilizer or use less and keep away from base of plants.
	Disease/damping off	Pasteurize soil (see page 171) for seedlings; use clean pots; improve drainage and avoid overwatering; add compost to soil.
Stunted, pale to yellow plants	Low fertility	Use foliar fertilizer now; test soil for nutrient balance for later crops.
	Acidic soil	Test pH and amend as needed.
	Poor drainage	Add organic matter and see chapter 7.
	Compacted soil	Break up compaction; see chapter 7.
	Insects or diseases	Inspect plants to identify; use appropriate control.
	Nematodes	Consult Cooperative Extension Service for help identifying; solarize soil (see page 160).
Stunted, purplish plants	Temperature too low	Plant at recommended time or use floating row covers (available at garden centers).
	Low phosphorus	Use foliar fertilizer; add bonemeal or other phosphorus source.
Spots or darkened areas on leaves and/or stems	Disease	Identify and use appropriate control; choose resistant varieties in future.
	Nutrient deficiency	See symptom chart in appendix.
	Fertilizer burn	Avoid getting fertilizer on leaves; dilute foliar types, following label directions.
	Chemical burn	Switch to organic controls; follow label directions; don't spray in hot sun.

Symptom	Possible Causes	Treatment/ Prevention
Brown leaf tips	Salt burn	Test soil salt levels (see chapter 7); flush salts from container plants.
	Low potassium, calcium	Test soil.
Wilting plants	Dry soil	Irrigate thoroughly.
	Poor drainage	Add organic matter and see chapter 7.
	Nematodes	Consult Cooperative Extension Service for help identifying; solarize soil (see page 160) .
	Wilt diseases	Remove infected plants; choose resistant varieties in future (and see chapter 7).
Weak and spindly plants	Too much nitrogen	Use less fertilizer next time.
	Too much water or poor drainage	Water less; add organic matter and see chapter 7.
	Too much shade	Move plants to sunnier spot.
	Plants too close	Thin; sow at recommended spacing.
No fruits form	Too much nitrogen	Use less fertilizer next time.
	Too hot or too cold	Plant at recommended time and/or use floating row covers.
	Nutrient deficiency	See chart in appendix.
Dry brown to black rot on blossom end of tomato	Low calcium	Add calcium source to soil.
	Extremely dry soil	Irrigate.
Abnormal growth and distorted leaves	Herbicide burn	Switch to other forms of weed control; follow label directions; don't use herbicide sprayers for other sprays.
	Viral diseases	Remove infected plants; choose resistant varieties; control insects that spread viruses, or grow plants under floating row covers.

Revised from *Down-to-Earth Gardening Know-How for the '90s,* by Dick Raymond (Storey Publishing, 1991)

Container Gardening

Some people grow plants in containers because they have no garden, others because they have extreme soil, and others just because they enjoy it. Whether indoors or out, plants in containers have slightly different requirements from those growing in the ground.

Providing Water and Drainage

For any container plants, drainage and watering are the main issues. Roots have nowhere to go if the soil is saturated or compacted. Poor drainage not only causes roots to starve for lack of oxygen, it also promotes rot and salt buildup. Even when drainage is adequate, more plants are killed by overwatering than by underwatering.

Container soils compact more easily than garden soils. Watering tends to settle the soil in a pot, and there aren't any earthworms or burrowing insects to loosen it up. Very fine commercial mixes used by some growers may work well for a short period but can easily become compacted in a matter of months.

To ensure long-term health, soils for containers must have a looser texture — more coarse sand and coarse organic matter — than garden soils. Perlite, a volcanic material heated to high temperatures to make it expand, is an excellent lightweight substitute for sand that provides even better drainage. Vermiculite (heat-expanded mica) is a good lightweight material for increasing the water-holding ability of container soils.

Keeping Containers Well Fed

Fertilizing is another important issue with container gardening. Overfertilizing causes salt buildup and other problems, but plant roots can't reach beyond the pot to search for nutrients if they're lacking. Since nutrients are easily washed away, more frequent feedings of light doses are better than infrequent full-strength doses.

Traditional liquid fertilizers supply only nitrogen, phosphorus, and potassium, so unless plants are frequently repotted into fresh soil they may suffer from deficiencies of calcium or micronutrients. To provide a fully balanced diet, include some compost in your soil mix and use liquid seaweed, fish emulsion, or compost tea instead of or alternating with synthetic fertilizers.

MASTER GARDENING TIP

Soil Polymers: Are They Worth the Cost?

Nontoxic, water-absorbing gels supposedly reduce the need to water plants. These gels are processed into small, beadlike crystals so you can mix a spoonful into the soil mix for each pot. No one argues that these polymers absorb water; a teaspoon can soak up a whole pint! The question is whether this water is then released for plant use.

Tests at the University of California at Riverside show that most of the water remains in the polymers. While there may be a slight increase in humidity (along the lines of setting houseplants on a water-filled tray of gravel), you'll still need to water frequently to keep your plants from drying out.

Polymers work best in pots of very coarse-textured (sandy) soils; in ordinary potting soil you might gain one extra day. But you might also find it harder to tell when plants need water.

Supplying Adequate Light

With houseplants, light is equally as important as drainage (and humidity and temperature aren't far behind). Most houseplants don't get enough. If you checked with a light meter, you'd see that light levels drop off dramatically within a few feet of any window. When plants that aren't getting quite enough light are pushed with lots of fertilizer, they produce weak growth that sends out "come get me" signals to aphids and other pests. Feed indoor plants just enough to encourage slow, healthy growth. (As a reward, you won't have to repot them as frequently.) In winter, cut back on feeding and watering to help plants slow their growth to match winter's reduced light levels.

HINT FOR SUCCESS

Watching Your Water

If your tap water is hard or very heavily chlorinated, try replacing it with other sources for at least part of the year. Set up a plastic trash can under a downspout during frost-free seasons to collect rainwater. If you use a dehumidifier, save the water it produces for your houseplants.

Controlling Salt Buildup

If you see a white crust on the soil of your houseplants, it's salt buildup. Too much salt can cause leaf tips and edges to turn brown. Salts are added by fertilizing and by using hard (alkaline) water. (Your water is hard if it doesn't produce good soapsuds and/or it requires water softeners.) Softened water contains the worst type of salt, sodium chloride. Never use it on your plants — it causes soil as well as growth problems.

Luckily, it's much easier to flush excess salts out of containers than out of the ground. The method is similar, though: Add enough water to wash away the excess salts. Water your plant(s) until water drains into and nearly fills the saucer below. Empty the saucer and repeat two or three times (more for small saucers). Treat several plants at once so you can let at least 20 minutes pass before repeating.

If a pot shows a heavy salt buildup, repot the plant in fresh soil. Soak the pot and scrub off the salt before replanting. Periodic flushing should keep salts from building up to a visible crust. You can also slow salt buildup by putting some gravel in the tray to lift the pot above any water that drains out.

A white crust on the soil of houseplants, or a ring where the soil meets the pot, is caused by excess salts in fertilizers and in some types of tap water.

Homemade Potting Soil

Commercial potting mixes vary in composition, pH, salt levels, nutrient balance, drainage, and how well they hold water. If you haven't had good luck with purchased mixes, try making your own.

Don't use soil straight from the garden. Even the best garden loams compact too quickly in containers. You must add coarse sand and lots of organic matter to counteract compaction and provide a good balance between drainage and water retention. Adding some soil is good — it supplies a wide array of nutrients. Just be sure to pasteurize soil (see opposite) before using it in mixes in order to kill any harmful organisms.

Recipes for Potting Soils

All the ingredients listed below are readily available at your local garden center. Don't use beach sand or road sand, as both could contain enough salt to harm plants. If you use compost, make sure it's finished (try the test on page 61). Clean all mixing tools, bowls, measures, and pots in hot, soapy water or a dishwasher before using. Mix all ingredients thoroughly.

Basic Container Soil Mix

1 part pasteurized soil (should be good, loamy soil; can substitute any commercial potting soil that's not a soilless mix)

1 part compost, leaf mold, shredded fir bark, or peat moss (the last two are acidic and low in nutrients, so compensate by adding 1 tablespoon, or 5 ml, bonemeal per quart — .95 ml — of finished mix)

1 part coarse sand or perlite (for cacti and other plants that require fast drainage, use 1 part sand *and* 1 part perlite)

Fortified Brooklyn Botanic Garden Mix

4 parts soil
2 parts sand
2 parts leaf mold or compost
1 part dried cow manure
½ cup (138 ml) bonemeal for every 2½ gallons (9.5 l) of mix

Cornell University Soilless Mix

8 quarts (7.6 l) fine horticultural vermiculite
8 quarts (7.6 l) shredded peat moss
2 tablespoons (30 ml) superphosphate (0-20-0)
2 tablespoons (30 ml) ground dolomitic limestone
8 tablespoons (118 ml) dried cow manure or steamed bonemeal

Seed-Germinating Mix

1 part fine commercial soilless mix
1 part perlite
1 part vermiculite

(Contains no nutrients, so feed seedlings periodically with diluted liquid fertilizer starting three weeks after they sprout.)

Pasteurizing Garden Soil

If you want to use garden soil in container mixes, pasteurize it first. You can do this in your oven. Pasteurizing kills the organisms that cause damping-off and other soilborne diseases. It's not the same as sterilizing, which kills all soil life. Don't pasteurize fresh compost, however, or you'll kill its disease-controlling organisms.

Materials

- Good garden soil
- Shallow roasting pan
- Oven roasting bag
- Meat thermometer
- Pot holders or oven mitts
- Timer

1 Cooked soil doesn't smell very good, so use an oven roasting bag to contain odors (and to keep soil from touching the pan). Dampen the soil thoroughly, so it will steam. Close the bag and poke a hole large enough to fit the stem of a meat thermometer. Don't let it get above 180°F (82°C).

2 Heat the soil to 160–170°F (71 to 77°C) for 30 minutes. Periodically insert thermometer into soil (avoid touching pan) to check temperature; try setting the oven to 200°F (93°C) until the soil reaches 165°F (74°C) and then lower the setting a bit. Turn it up again if the soil temperature drops below 155°F (68°C). Make up for any time the soil measured below the desired range by adding that period to the timer.

3 The baked soil is ready to use in a mix as soon as it's cool enough to handle. Let it cool completely before placing seeds or plants in it.

Managing Soils for Annual Flowers

Most annual flowers prefer the same general conditions that vegetables do, except they need less fertilizer. The general strategies for soil care are the same (unless you're growing annuals mixed in among perennials, in which case use the strategies for perennials). Like vegetable beds, annual beds can be emptied at the end of the season (or between successive plantings in mild climates). That makes it convenient to till in chopped leaves, spread compost, or grow a winter cover crop to supply organic matter.

Adding Nutrients

Most annuals are average feeders. Modern varieties are bred to withstand neglect, but they'll perform best in well-drained soil that's amended with some organic matter. Many will bloom well with only compost as fertilizer, assuming your soil's overall nutrient balance is good. A few annuals are heavy feeders; these need soil enriched with ample compost or aged manure plus a midseason sidedressing to keep blooming well. A few others are light feeders; these bloom best in average to somewhat poor soil with no added fertilizer, just a little compost. (See the chart opposite to learn which flowers like light or heavy diets.)

A good rule of thumb is to use half as much fertilizer for annuals as you would for vegetables. If you use liquid fertilizers, mix them at half strength. When choosing a general-purpose blend, opt for one with less nitrogen, such as 5-10-10. If you give flowering annuals too much nitrogen, you'll end up with lots of lush leaves and few blooms.

For average feeders, mixing a slow-release synthetic formula or an organic fertilizer into the bed before planting should supply enough nutrients for the season. If you didn't add nutrients before planting, fertilize about a month after planting or when blooms appear. If plants appear to be slowing down (and it's not because you forgot to clip off faded blooms), try a boost of liquid seaweed diluted to half strength or compost tea (but not manure tea). Sidedress heavy feeders again about midway through the season.

MASTER GARDENING TIP

Weed Control for Annuals

While many gardeners don't think of cover crops in annual beds, winter rye or annual ryegrass can be a great ally in controlling weeds the following season. Both work best after annuals are killed by frost; growing them under floating row covers allows you to sow them a bit later than normal. Just make sure to dig either under three or four weeks before planting seeds or transplants.

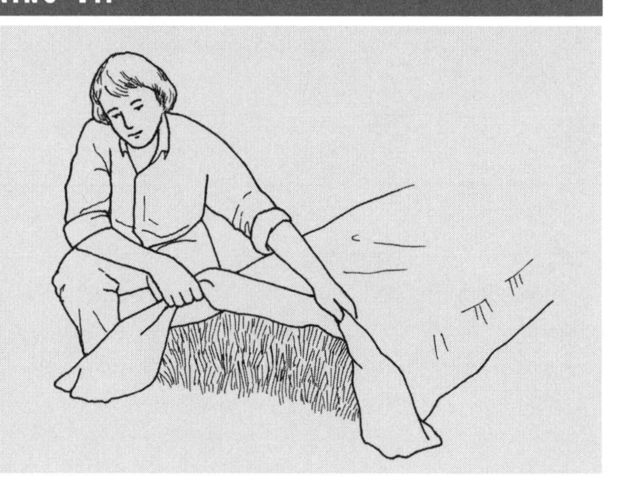

In mild areas with long growing seasons, you may need to fertilize again after a couple of months to prolong the display. In very sandy soils, remember to divide the recommended amount of fertilizer into smaller portions and apply it more frequently throughout the season.

Watering and Mulching

Since annuals don't have time to develop extensive roots capable of reaching far for water, they depend on you to supply water when rains are lacking. Where rains are unreliable, a good organic mulch around your annuals saves you a lot of time while reducing plants' heat and water stress. Mulch also prevents the soil from crusting, which in turn improves aeration and water penetration.

Caring for Container Plants

Annuals in containers need watering and fertilizing more often than those in the ground. They're usually crowded much more closely together and have a limited supply of soil within reach. Also, nutrients are washed away whenever water runs out the bottom of the container. Don't give container plants a stronger dose, just feed them more often.

Feed annuals in containers with a liquid fertilizer such as a seaweed/fish emulsion blend at half strength every two to three weeks. Adjust the frequency by watching how your annuals respond. If you use a concentrated formula such as 15-30-15, dilute it to one-quarter strength.

MASTER GARDENING TIPS

Annuals That Require Special Diets

Heavy feeders

- Begonias, wax or tuberous (*Begonia* spp.)
- Butterfly flower (*Schizanthus* x *wisetonensis*)
- Caladium (*Caladium* x *hortulanum*)
- Canna (*Canna* x *generalis*)
- Coleus (*Coleus* x *hybridus*)
- Dahlias (*Dahlia* hybrids)
- Heliotrope (*Heliotropium arborescens*)
- Impatiens (*Impatiens* hybrids)
- Larkspur (*Consolida ambigua*)
- Monkey flower (*Mimulus* x *hybridus*)
- Ornamental kale (*Brassica oleracia*)
- Polka-dot plant (*Hypoestes phyllostachya*)
- Snapdragon (*Antirrhinum majus*)
- Sweet pea (*Lathyrus odoratus*)
- Torenia (*Torenia fournieri*)

Light feeders

- Amaranths (*Amaranthus* spp.)
- Calendula (needs even moisture) (*Calendula officinalis*)
- Chinese forget-me-not (*Cynoglossum amabile*)
- Cleome (*Cleome hasslerana*)
- Corn cockle (*Agrostemma githago*)
- Gazanias (*Gazania* spp.)
- Iceland poppy (*Papaver nudicaule*)
- Morning-glory (*Ipomoea tricolor*)
- Nasturtium (*Tropaeolum majus*)
- Painted-tongue (*Salpiglossis sinuata*)
- Shirley poppy (*Papaver rhoeas*)

Managing Soils for Perennial Flowers

Perennials like the same soil as most other plants: well drained, evenly moist, enriched with ample organic matter, and with a good balance of nutrients. Since it's not as easy to build the soil around existing plants, it pays to prepare beds well before planting.

HINT FOR SUCCESS

Replenishing Organic Matter

Since you can't grow a cover crop or turn under chopped leaves, topdressing and mulching are your options for replenishing organic matter in existing perennial beds. Try to spread an inch of compost or aged manure over the bed, or at least around every plant, each year. Unless slugs, snails, and/or dampness are major problems, keep plants mulched to maintain even soil moisture, keep down weeds, and help supply organic matter.

Fertilizing

Light feeders should get only compost each year. Average and heavy feeders will benefit from a little fertilizer. Avoid too much nitrogen, though; you'll get floppy plants that require staking, as well as fewer flowers. (Light feeders growing in rich soil will also get floppy.) Scratch granules into the soil around plants, working carefully to avoid disturbing roots. Covering with an inch of compost has the same effect as mixing fertilizer an inch deep. Or use a liquid fertilizer.

As soon as new growth appears in spring, feed with a balanced, lower-nitrogen fertilizer such as 2-4-2 or 5-10-5. Slow-release and organic formulas are best — smaller amounts are released steadily over a longer period of time. If the label gives no specific recommendations for flowers, use at half the rate recommended for vegetables. Use either foliar feeding or sidedressing to give heavy feeders a midseason boost.

Perennials with an exceptionally long bloom season (over six weeks) benefit from an additional boost. Give them a foliar feeding of compost tea, half-strength liquid seaweed, or one-quarter-strength balanced liquid fertilizer every two to three weeks through their blooming period.

MASTER GARDENING TIP

Soil Care for Bulbs

Good drainage is even more essential for bulbs than it is for perennials. It's the most important factor to ensure many years of reliable blooms.

Most bulbs respond well to annual feedings of compost and no additional fertilizer. If you want to add fertilizer, consider one of the special formulas for bulbs. The most common of these is a synthetic blend with a formula of 9-9-6, especially good for tulips, which contains nitrogen in a slow-release form. A good organic bulb fertilizer with a formula of 5-10-12 is also available. Daffodils prefer 5-10-20, but this formula is hard to find. Bonemeal isn't the best bulb fertilizer because it may attract animals. Also, research has shown that, as processed today, bonemeal is relatively low in most nutrients.

If you do add fertilizer when planting, make sure it's either organic or a slow-release formula to avoid burning bulb roots. Mix it into the soil below bulbs and cover with at least an inch of unfertilized soil.

Fall is the best time for feeding existing plantings. Use a slow-release or organic formula and scratch it into the soil (or cover with a 1-inch topdressing of compost) to keep it from washing away. Summer-blooming bulbs, such as lilies, should instead be fertilized as soon as their shoots emerge in spring.

Soil Needs of Some Perennials

Heavy Feeders

Bearded iris* (*Iris* spp.) oo
Bee balm* *(Monarda didyma)* oooo
Cardinal flower* *(Lobelia cardinalis)* ooooo
Chrysanthemums* (*Dendranthema* spp.) ooo
Clematis (*Clematis* spp.) ooo
Delphinium *(Delphinium)* ooo
Lilies (*Lilium* spp.) ooo
Monkshoods (*Aconitum* spp.) oooo
Peony (*Paeonia* spp.) ooo
Phlox* *(Phlox paniculata, P. maculata)* ooo
Tree peony *(Paeonia suffruticosa)* ooo
Turtleheads* (*Chelone* spp.) oooo

**Plants that don't require another feeding but respond well to one*

Average Feeders

Alliums (*Allium* spp.) ooo
Asters (*Aster* spp.) oooo
Baby's breaths (*Gypsophila* spp.) ooo
Balloon flower *(Platycodon grandiflorus)* ooo
Baptisias (*Baptisia* spp.) ooo
Bellflowers (*Campanula* spp.) ooo
Bergenias (*Bergenia* spp.) oooo
Butterfly weed *(Asclepias tuberosa)* o
Catmints (*Nepeta* spp.) oo
Columbines (*Aquilegia* spp.) oo
Coral bell *(Heuchera sanguinea)* ooo
Coreopsis (*Coreopsis* spp.) oo
Cranesbills (*Geranium* spp.) ooo
Daylilies (*Hemerocallis* spp.) oo
Ferns (many species) oooo
Fleabane *(Erigeron speciosus)* ooo
Foxgloves (*Digitalis* spp.) ooo
Gayfeathers (*Liatris* spp.) oo
Helenium *(Helenium autumnale)* oooo
Hostas (*Hosta* spp.) oo
Japanese anemone *(Anemone japonica)* oooo
Lady's mantle *(Alchemilla mollis)* ooo
Lamium *(Lamium maculatum)*oo
Lily-of-the-valley *(Convallaria majalis)* oooo
Lungworts (*Pulmonaria* spp.) oooo
Lupines (*Lupinus* spp.) ooo
Mallows (*Malva* spp.) ooo
Pincushion flowers (*Scabiosa* spp.) oo
Pinks (*Dianthus* spp.) oo
Poppies (*Papaver* spp.) ooo
Red-hot-poker *(Kniphofia uvarium)* ooo
Solomon's seals (*Polygonatum* spp.) oooo
Spiderwort *(Tradescantia* x *andersoniana)* ooo
Stokes' aster *(Stokesia laevis)* oo
Sunflowers (*Helianthus* spp.) ooo
Veronicas (*Veronica* spp.) oo

Light Feeders

Artemisias (*Artemisia* spp.) oo
Basket-of-gold *(Aurinia saxatilis)* o
Black-eyed Susan *(Rudbeckia fulgida)* ooo
Blanket flowers (*Gaillardia* spp.) o
Candytuft *(Iberis sempervirens)* ooo
Cinquefoils (*Potentilla* spp.) oo
Epimediums (*Epimedium* spp.) ooo
Evening primroses (*Oenothera* spp.) oo
Golden marguerite *(Anthemis tinctoria)* o
Jupiter's beard *(Centranthus ruber)* oo
Lamb's-ear *(Stachys lanata)* ooo
Obedient plant *(Physostegia virginiana)* oooo
Prickly pear *(Opuntia humifusa)* oo
Purple coneflower *(Echinacea purpurea)* oo
Rock cresses (*Aubretia, Arabis* spp.) oo
Russian sage *(Perovskia atriplicifolia)* oo
Salvias (*Salvia* spp.) ooo
Sea hollies (*Eryngium* spp.) o
Snow-in-summer *(Cerastium tomentosum)* oo
Soapworts (*Saponaria* spp.) oo
Stonecrops (*Sedum* spp.) ooo
Tansy *(Tanacetum vulgare)* oo
Thrifts (*Armeria* spp.) o
Verbenas (*Verbena* spp.) o
Yarrows (*Achillea* spp.) oo
Yuccas (*Yucca* spp.) o

o = very well-drained (dries quickly)
oo = moderate to well-drained
ooo = moderately well-drained
oooo = moist to well-drained
ooooo = constantly moist (not wet)

Managing Soils for Lawns

Since most lawn grasses are under constant stress from less-than-ideal soils, lawns respond well to soil-building efforts. Grass in compacted soil that's low in organic matter will build up thatch, because there are few earthworms and other organisms to break down the thatch. Once it gets over ¾ inch thick, thatch blocks air circulation to the soil and keeps water from soaking in, further reducing populations of soil organisms. Eventually, water may puddle on lawns, causing disease problems.

Lawns with compacted soil are more prone to pests. Treating these symptoms rather than the underlying cause — unhealthy soil — may just intensify the problem. Pesticides kill soil organisms as well as pests, but killing soil organisms impairs soil health and therefore makes grass even more susceptible to insects and diseases.

Restoring Soil Life and Organic Matter

The first steps to a lush, low-maintenance lawn are hard to separate — restore soil life and restore organic matter levels. To restore soil life, create the same conditions that promote good grass growth: soil that's slightly acidic (pH 6.5) with good drainage, good aeration, and a good nutrient balance. Test your soil to see which nutrients are out of balance and add only those nutrients that are lacking. Fix drainage problems (see chapter 7). Rent an aerator (power lawn corer) to increase aeration for the short term; adding organic matter and restoring soil life will help in the long term.

When to Start Over

If your grass is really thin or weeds are really thick, it may be faster to start all over again. Two plantings of buckwheat (or buckwheat and rye) will get rid of weeds while adding lots of organic matter. If you don't want to go the green manure route, turn a 1- to 2-inch layer of organic matter into the top 4 or 5 inches of soil before reseeding. Check for underlying hardpan and use a broadfork if you find any.

DID YOU KNOW?

How to Mistreat a Lawn

The soil around houses is often mistreated long before the homeowner arrives. Topsoil is usually removed during house construction, to be sold for a good price with just a little remaining pushed to one side. Building codes may require compaction of soil around the foundation to reduce settling of the finished house. Bulldozers and trucks contribute more compaction. No wonder grass has a hard time growing in the thin topsoil spread after construction!

Ordinary wear and tear also compact soils. Kids cut across the lawn or play ball on it, and people drive over it. Even mowing can compact wet, clay soils. Topdressing with organic matter, once standard practice, went out of fashion as special lawn fertilizers became popular. Removing grass clippings combined with lack of topdressing causes gradual depletion of organic matter, which just makes soils more susceptible to compaction. To avoid mistreating your lawn, follow the suggestions at right.

Including white clover in your lawn can reduce nitrogen needs by up to one-third. (Remember to use the right inoculant.)

Mowing

While healthy soil greatly increases your grass's ability to deal with stresses such as drought, disease, or nutrient imbalance, good management can minimize those stresses for even better-looking lawns. Instead of removing grass clippings, leave them on the lawn to replenish the soil's organic matter and nutrients. Invest in a mulching mower. These chop grass finely, returning all those nutrients to the soil quickly without making your lawn look like a hayfield.

Grass can tolerate stress better when it is long, so keep your grass 2 to 3 inches (5 to 8 cm) tall. (The chart on page 179 notes a few species that should be kept shorter.) Mowing stresses plants, so try to remove only one-third of the blade's height at one time. If you miss a mowing or two, cut off a third and wait a few days before cutting again.

Watering and Fertilizing

Frequent, light watering just makes grass more susceptible to drought by encouraging short roots. Water deeply and infrequently to stimulate deep, drought-resistant roots. Run the sprinkler until water starts to run off, turn it off for at least an hour to let the water soak in deeply, and turn it on again until water starts to run off.

Don't overfertilize. Lush, soft grass is more vulnerable to disease and can't handle drought. Extra fertilizer suppresses the activity of soil organisms and leads to thatch buildup. (It also means you have to mow more often.) Test your soil periodically to see which nutrients it really needs and whether it really needs lime before adding anything. To feed steadily over a longer period of time, choose organic and/or slow-release nutrient sources. Using a mulching mower and topdressing will greatly reduce your fertilizer needs after a couple of years.

Compost Is Great for Lawns

A U. S. Department of Agriculture study found that topdressing lawns with composted sewage sludge furnished many benefits not supplied by ordinary lawn fertilizers. The compost topdressing reduced soil compaction and increased both water-holding capacity and air exchange. Applying it at a rate of 3,300 to 6,600 pounds per 1,000 square feet (180 to 360 metric tons per hectare) improved the root environment for better grass growth.

In addition to adding low to moderate levels of nutrients, compost increases both pH and pH buffering capacity. It stimulates soil organisms and increases overall pest and disease resistance. Some organic lawn fertilizers are now inoculated with some of the specific disease-fighting microorganisms found in good compost.

While the study used composted sewage sludge, any good compost will supply the same benefits. If you have a municipal composting facility nearby, ask about its compost. See what's available at your local garden center; composted cow manure can be found almost everywhere. Commercial compost is usually fine enough to apply with a fertilizer spreader. Homemade compost is great; sift it to remove large pieces before using it on the lawn (see page 80).

MASTER GARDENING TIPS

Regional Tips for Lawn Care

Warm, arid climates (Southwest U.S.)

- Fertilize cool-season lawn grasses in October or November. (If cool-season grasses are irrigated year-round, fertilize again in May or June.)
- Fertilize warm-season lawn grasses in April or May and again in August. (If lawns are irrigated all summer, apply smaller doses monthly from May through August.)

Warm, humid climates (Southeast U.S.)

- Fertilize lawns in June and again in August (or divide recommended amounts by four and apply one-quarter each month from May through August). If lawns are overseeded with annual ryegrass, feed these in September or October.

Cool climates (central and northern U.S.)

- Fertilize lawns in spring once grass starts growing, and again in early fall.

Topdressing a Lawn

Topdressing with organic matter is the most effective way to reduce thatch in lawns. It increases microbe populations and activity by providing the food they need. Large microbe populations not only improve soil health, they also break down thatch quickly and steadily, preventing a gradual buildup. Compost is the best topdressing, but you can use any fine material, such as dehydrated manure. (Don't use peat moss, which repels water.)

The process is basically the same as spreading fertilizer or lime, though you're using more material. Aim for a layer about ¼ inch (.6 cm) deep. Since the material you're spreading is a low-level and slow-release source of nutrients, you don't have to worry about creating a striped lawn. You can topdress lawns at any time, but spring and fall are best. Fall is particularly good if you're going to overseed sparse lawns with one of the improved varieties or blends of grass seed. Spread the seed and then cover with the topdressing to improve germination. *(For metric equivalents, see "Useful Conversions" on page 208.)*

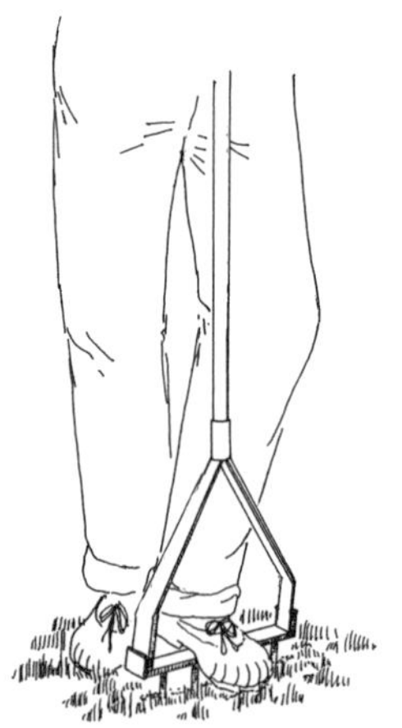

1 If the thatch on your lawn is more than ¾ inch thick, rake the lawn first to remove it. If you're planning to aerate your lawn, do it before topdressing so that some of the topdressing can sift down into the holes left by the lawn corer.

2 Calculate how much organic matter you'll need — about 23 cubic feet for every 1,000 square feet of lawn. It must be fairly fine in order to disappear between grass blades. If you're using homemade or coarse commercial compost, sift it through a ¼- or ½-inch mesh.

3 If your topdressing is very fine and/or granulated, you can spread it ¼ inch deep with a fertilizer spreader. Otherwise, distribute several small piles evenly over your lawn. Rake out the piles into a roughly ¼-inch-deep layer.

Choosing the Right Grass

To get the best-looking lawn, choose a mix of grasses suited to your conditions. A mix of two or three different grasses, or at least two or three varieties of the same grass, increases your lawn's tolerance for heat, drought, and pest problems. New varieties of the most common grasses have been bred to offer much greater disease resistance; some even resist insects.

Choose cool-season types for the northern half of the country and higher altitudes. Warm-season grasses withstand heat but go dormant (turn brown) in cool or cold winters; they need much less water than most cool-season types. If you want to spend less time on your lawn, choose species and varieties with a greater tolerance for drought and low fertility.

Types of Lawn Grasses

Type	Preferred pH	Comments
Cool-Season		
Bentgrass	5.5–6.5	Low tolerance to heat, drought, traffic, and low fertility; keep short (1 in., or 2.5 cm); prone to thatch buildup.
Bluegrass, Kentucky	6.0–7.5	Fairly adaptable; prefers full sun and cooler summers; many disease-resistant varieties available.
Fescues, fine	5.5–7.5	Adaptable; tolerate low fertility, shade, and drought; several disease-resistant varieties available.
Fescues, tall	5.5–7.5	Adaptable; tolerate heat, drought, low fertility, and traffic; rarely form thatch; best mixed with other species.
Redtop	5.0–7.5	Coarser, more tolerant form of bentgrass; keep short (1 in., or 2.5 cm).
Ryegrass, perennial	5.0–8.0	Good for high-traffic areas; good for new lawns (establishes quickly); several disease- and insect-resistant varieties available.
White clover	5.5–7.0	Adaptable; drought tolerant; can supply up to 30 percent of lawn's nitrogen needs (inoculate when planting); good mixed with lawn grasses.
Warm-Season		
Bermuda grass	5.0–7.0	Fine texture; adaptable; tolerates heat, cold, drought, and traffic; keep short (summer 1 in., or 2.5 cm; spring and fall 1½ in., or 4 cm); prone to thatch; look for newer varieties ("improved Bermuda grass").
Buffalo grass	6.0–8.5	Very drought tolerant (needs only one-fifth as much water as Kentucky bluegrass); good traffic tolerance; new shorter varieties need little or no mowing.
Carpet grass	4.5–7.0	Somewhat coarse; tolerates low fertility; little tendency to form thatch; low tolerance to drought and traffic.
Centipede grass	4.0–6.0	Somewhat coarse; tolerates drought and somewhat low fertility; best for low-traffic areas.
Grama grass, blue	6.0–8.5	Tolerates drought, heat, and cold but not low fertility; can mix with buffalo grass.
St. Augustine grass	6.0–8.0	Somewhat coarse; tolerates shade and heat; prone to thatch; needs warm climate.
Zoysia grass	4.5–7.5	Fine texture; adaptable; tolerates heat, cold, drought, and heavy traffic; keep short (summer 1 in., or 2.5 cm; spring and fall 1½ in., or 4 cm).

Managing Soils for Woody Ornamentals

Soil care for ornamental trees, shrubs, and vines is essentially the same as for fruit and nut trees. You're investing in more expensive plants that can't be moved once they mature, so it pays to treat them well. Investigate potential problems with drainage and pH before you buy plants so you have plenty of time (ideally six months or more) to make any corrections. You won't be able to correct drainage after you plant. Also remember that any soil improvement should extend over a large area. When soil is improved in just the planting hole, pampered roots won't venture out into the unimproved soil beyond. To encourage roots to venture out, make planting holes wider at the top than at the bottom and roughen up the sides.

Fertilizing

Most trees, shrubs, and vines aren't heavy feeders, so an annual topdressing of compost will keep them happy. Trees growing in lawns get fed automatically every time you topdress or fertilize the lawn. New transplants may appreciate some fertilizer the first few years, until their root systems grow large enough for them to forage for nutrients.

Heavy Feeders

Some of the flowering ornamentals are heavy feeders, especially hybrids and those bred for abundant blooms or a long season. Feed deciduous types such as roses and clematis when new growth appears in the spring. If they are rebloomers, feed again as each flush of blooms ends to encourage more blooms (but stop by midsummer where winters are cold). Feed flowering evergreens such as azaleas, camellias, and rhododendrons immediately after bloom and again in mid- to late summer. If you've built up your soil and are giving these plants compost every year, you'll need less fertilizer than the labels recommend.

Good drainage and balanced nutrients will help woody plants survive harsh winters.

Increasing Winter Hardiness

Two soil characteristics can have a great effect on the winter hardiness of trees and shrubs: drainage and nutrient balance. Poor drainage is more apt to kill dormant and evergreen woody plants than temperature extremes. A combination of both is especially deadly.

One nutrient, nitrogen, reduces winter hardiness while another, potassium, increases it. Too much nitrogen causes lush, tender growth that's more easily damaged by cold. If you live where winters are cold or have any trouble with winter hardiness, stop fertilizing all ornamentals (even heavy feeders such as roses) by midsummer. This gives plants plenty of time to harden off tender, new growth before cold temperatures arrive.

Making sure your soil has adequate levels of potassium will improve overall winter hardiness, especially for newly transplanted trees and shrubs. (Phosphorus helps, too.) Unless soil tests show higher than medium levels, add a little greensand as a good slow-release source of this nutrient. You won't need to reapply it for several years. Give new plants a short-term boost with a foliar feeding of liquid seaweed late in the growing season; seaweed can increase winter survival of woody ornamentals.

Watching Soil pH for Acid-Lovers

Acid-loving evergreens will be quick to let you know when their nutrient balance is off by developing yellow leaves (chlorosis). The best way to prevent this is to test the soil pH periodically and amend as needed before any symptoms show. Use acidic mulches such as pine needles and acidifying fertilizers to help keep soil acidic.

MASTER GARDENING TIP

Don't Overlook Native Trees and Shrubs

Try to choose ornamentals that prefer the conditions your garden has to offer. You'll be rewarded with better looks, fewer pest problems, and in dry climates you won't have to water as often. If all or part of your yard contains difficult soils, you also won't have to work so hard to change them. When a plant is naturally adapted to your soil and climate, or to very similar conditions elsewhere, it has a head start over plants adapted to different soils and climates.

Books and entire nurseries specializing in native woody plants are much more common now than they were a generation ago, so it's much easier to find information as well as plants. You may be able to get design ideas, as well as information on sources, from public gardens near you.

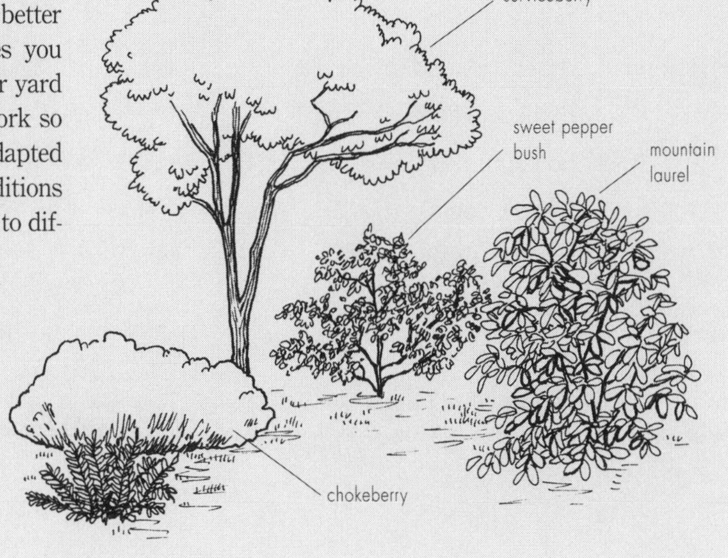

Suggested Woody Plants for Various Soil Types

Most shrubs prefer the same evenly moist, relatively fertile, well-drained, slightly acidic soil that other plants like. If you've got such soil, you can grow many common shrubs (such as roses) that aren't listed here. These are alternatives for more challenging conditions. Ask your Cooperative Extension Service or local landscape professionals about varieties and species particularly suited to your area.

Soil Type	Shrubs	Trees	Perennial Vines
Acidic	Azaleas and rhododendrons (*Rhododendron* spp.) Camellias (*Camellia* spp.) Enkianthus *(Enkianthus campanulatus)* Gardenia *(Gardenia jasminoides)* Heaths and heathers (*Erica, Calluna* spp.) Japanese pieris *(Pieris japonica)* Leucothoes (*Leucothoe* spp.) Mahonias (*Mahonia* spp.) Mexican orange *(Choisya ternata)* Mountain laurel *(Kalmia latifolia)*	Australian tea tree *(Leptospermum laevigatum)* Hemlocks (*Tsuga* spp.) Magnolias (*Magnolia* spp.) Oaks (*Quercus* spp.) Pines (*Pinus* spp.) Sassafras *(Sassafras albidum)* Shadbushes (*Amelanchier* spp.) Sourwood *(Oxydendron arboreum)* Spruces (*Picea* spp.) Stewartias (*Stewartia* spp.)	Trumpet vine *(Campsis radicans)*
Alkaline	Acacias (*Acacia* spp.) Barberries (*Berberis* spp.) Bottlebrushes (*Callistemon* spp.) Butterfly bushes (*Buddleia* spp.) Cypress (*Cupressus* spp.) Daphnes (*Daphne* spp.) Forsythias (*Forsythia* spp.) Junipers (*Juniperus* spp.) Pyracanthas (*Pyracantha* spp.)	Black locust *(Robinia pseudoacacia)* Bur oak *(Quercus macrocarpa)* Catalpas (*Catalpa* spp.) Flowering ash *(Fraxinus ornus)* (and most ashes) Golden rain tree *(Koelreuteria paniculata)* Honey locust *(Gleditsia triacanthos)* Sycamores (*Platanus* spp.)	Boston ivy *(Parthenocissus tricuspidata)* Clematis (*Clematis* spp.) Hardy kiwi *(Actinidia arguta)* Passionflowers (*Passiflora* spp.)

Soil Type	Shrubs	Trees	Perennial Vines
Wet	American arborvitae *(Thuja occidentalis)* American elder *(Sambucus canadensis)* Carolina allspice *(Calycanthus floridus)* Chokeberry *(Aronia arbutifolia)* Clethra *(Clethra alnifolia)* Cranberry *(Vaccinium macrocarpon)* Red cap eucalyptus *(Eucalyptus erythrocorys)* Red osier dogwood *(Cornus sericea)* Salal *(Gaultheria shallon)* Spice bush *(Lindera benzoin)* Many viburnums (*Viburnum* spp.) Winterberry *(Ilex verticillata)*	Alders (*Alnus* spp.) Bald cypress *(Taxodium distichum)* Black tupelo *(Nyssa sylvatica)* Coast redwood *(Sequoia sempervirens)* Green ash *(Fraxinus pennsylvanica)* Larches (*Larix* spp.) Pin oak *(Quercus palustris)* Poplars (*Populus* spp.) Red maple *(Acer rubrum)* River birch *(Betula nigra)* Sweet gum *(Liquidambar styraciflua)* Willows (*Salix* spp.)	Virginia creeper *(Parthenocissus quinquifolia)*
Poor*	Bush cinquefoils (*Potentilla* spp.) Cotoneasters (*Cotoneaster* spp.) Junipers (*Juniperus* spp.) Manzanitas (*Arctostaphylos* spp.) Oleander *(Nerium oleander)* Rock roses (*Cistus* spp.) Most sumacs (*Rhus* spp.) Warminster broom *(Cytisus praecox)* Most shrubby herbs	Acacias (*Acacia* spp.) Cottonwoods (*Populus* spp.) Eucalyptus (*Eucalyptus* spp.) Japanese pagoda tree *(Sophora japonica)* Mimosa *(Albizia julibrissin)* Russian olive *(Elaeagnus angustifolia)*	Crimson glory vine *(Vitis coignetiae)* English ivy *(Hedera helix)* Trumpet vine *(Campsis radicans)* Wisterias (once established) (*Wisteria* spp.)
Sandy/ Dry**	Amur maple *(Acer ginnala)* Bayberries (*Myrica* spp.) Beauty-bush *(Kolkwitzia amabilis)* Ceanothus (*Ceanothus* spp.) Chinese photinia *(Photinia serrulata)* Japanese privet *(Ligustrum japonicum)* Junipers (*Juniperus* spp.) Rugosa rose *(Rosa rugosa)* Sumacs (*Rhus* spp.)	Ailanthus *(Ailanthus altissima)* Eastern red cedar *(Juniperus virginiana)* Ginkgo *(Ginkgo biloba)* Gray birch *(Betula populifolia)* Japanese black pine *(Pinus thunbergiana)* Mugho pine *(Pinus mugo* var. *mugo)* Osage-orange *(Maclura pomifera)* Sassafras *(Sassafras albidum)* Siberian elm *(Ulmus pumila)*	English ivy *(Hedera helix)* Grapes (*Vitis* spp.) Persian ivy *(Hedera colchica)*

* *adequate drainage*

** *plants may need regular watering until established*

Managing Soils for Tree Fruits and Nuts

Good soil care is even more important for fruit and nut trees than it is for vegetables or bush fruits. The plants are more expensive, and they also live much longer. Poor soil can be costly in terms of time as well as money. On the brighter side, with these plants you reap greater benefits from good soil care over a longer time.

Preparing the Soil

Test your soil and amend to correct pH problems six months ahead of planting for the best results. Unless your soil is rich in phosphorus, it's a good idea to mix in a slow-release form (rock phosphate) at the same time for good root growth. Find out the expected diameter of your tree when mature. (Nut trees get very large!) Try to amend in a circle of roughly the same diameter (or as large as you can manage).

When planting several trees, incorporate organic matter over the entire area. Growing and tilling under a cover crop the previous season is a great way to do this. Or spread organic matter over the entire area and dig it into the soil. Either will encourage good root growth throughout the area. New research shows that if soil is improved just in the planting hole, pampered roots won't venture out into the unimproved soil beyond.

HINT FOR SUCCESS

Err on the Light Side

Take fertilizer recommendations with a grain of salt — they may not apply to your type of soil. For specific advice, ask local growers, your state university, or your Cooperative Extension Service.

If trees put on less than a few inches of new growth each year (or show pale leaves from nitrogen deficiency), slowly increase the amount or frequency of feeding. If growth is too lush, yields are poor, or leaves are dark green from lots of nitrogen, cut back.

Organic Matter and Mulch

After planting, increase humus levels gradually and indirectly by spreading organic matter on the soil surface. Add an inch (2.5 cm) or more of compost at least once a year and maintain a thick layer of organic mulch year-round. Earthworms will gradually work this organic matter into the soil for you.

Whatever your soil, remember that roots grow out at least as far as the outermost branches. (Roots can grow up to three times as far for some trees in good soil!) If possible, keep the soil surface mulched out to the farthest branches. When you spread fertilizer or compost, it should ideally cover this entire area as well. There's no need to spread fertilizer within a foot of the trunks of mature trees, as few feeder roots grow there.

Fertilizing

Feed trees in spring after the ground thaws. Once a year is enough for mature trees. Young trees (those not old enough to bear fruit) benefit from a second feeding in June. In mild climates, feed evergreens such as citrus three times a year — late winter, June, and August — giving one-third the total recommended amount each time. In cold climates, don't fertilize after midsummer: Late, lush growth increases the chance of winter damage.

Use an all-purpose, balanced fertilizer. Start with the rates recommended on the label and adjust in subsequent years to match the growth of your trees. If you grow white clover under your trees, switch to a formula with no nitrogen. This green manure supplies all the nitrogen fruit trees need. Once trees are old enough to bear, too much nitrogen leads to watery fruit of poor quality. Feeding with compost offers many benefits — including allowing you to use less fertilizer than the recommended rates.

Soil Preferences of Tree Fruits and Nuts

Tree	Preferred pH*	Preferred Soil
Almond	6.0–7.0	Rich, well-drained soils
Apple	5.0–6.5	Well-drained loam
Apricot	6.0–7.0	Rich, light loam; tolerates most soils if well drained
Avocado	6.0–8.0	Deep, rich, very well-drained soils; low salt tolerance (subject to salt burn)
Cherry, sour	6.0–7.0	Tolerates heavier soils than sweet cherries
Cherry, sweet	6.0–7.5	Light, sandy soils (but needs regular watering)
Crab apple	6.0–7.5	Well-drained loam
Filbert/hazelnut	6.0–7.0	Deep, well-drained loam or sandy loam
Grapefruit	6.0–7.5	Well-drained, evenly moist loam
Hickory	6.0–7.0	Most soils if deep and well drained
Kumquat	5.5–6.5	Well-drained, evenly moist loam
Lemon	6.0–7.0	Well-drained, evenly moist loam
Olive	5.5–6.5	Most well-drained soils; tolerates drought and alkaline, shallow, or stony soils; benefits from occasional nitrogen, but needs no fertilizer to fruit on average soils
Orange	6.0–7.5	Well-drained, evenly moist loam
Peach and nectarine	6.0–7.5	Well-drained, light soils (because these warm up quickly in spring); in cold climates go light on fertilizer to reduce winter damage
Pear	6.0–7.5	Loamy to heavy soils; tolerates poor drainage and damp soils (unlike most fruit trees)
Pecan	6.4–8.0	Well-drained, deep soils (6 to 10 ft., or 1.8 to 2.9 m, deep); very low salt tolerance; zinc deficiency causes oddly bunched twigs
Plum	6.0–8.0	Rich, well-drained soils with abundant organic matter; European and Damson types prefer clay or heavy loam, but Japanese types prefer lighter soils
Walnut	6.0–8.0	Well-drained, deep, rich soils

** Most plants will tolerate a wider pH range. Citrus, for example, can tolerate a pH of 8 but may need micronutrients supplied in foliar sprays or chelates.*

If you have drainage problems, you'll have to raise the soil level to improve growing conditions for fruit and nut trees. Create a gently sloping berm at least 4 feet in diameter (3 feet for very dwarf trees) and 4 to 6 inches higher at the center. Using a broadfork over a large area around the planting hole can also help (see page 143).

Herbs Like It Lean

Many herbs are native to the eroded hillsides around the Mediterranean, so it's not surprising that they like lean, well-drained soil. You don't have to give herbs a starvation diet, though, to get good results. Ordinary garden soil with average fertility is fine if you leave off the fertilizer. Lots of nitrogen produces lush-looking herb leaves, but it reduces the concentration of the essential oils that give herbs their aroma and flavor (and some of their medicinal properties). If you want your herbs to taste good, hold the nitrogen.

Most herbs aren't fussy about soil acidity and grow well in slightly acidic to slightly alkaline soil. A pH of 6.5 to 6.8 is ideal. If your soil produces good vegetables, you shouldn't need to adjust the pH.

> **MASTER GARDENING TIPS**
>
> **Homes for Your Herbs**
>
> - Integrate your annual herbs into your vegetable beds.
> - Perennial herbs are easier to manage if treated like asparagus or rhubarb. Grow perennials together in their own bed to simplify soil care in the vegetable bed.
> - A slight slope is a great spot for a few herbs because it offers good drainage; terrace it or design an herbal rock garden.

Feeding with Compost

For many herbs, an inch or so of compost supplies all the nutrients they need. Those herbs that like a slightly richer diet are listed in the box on the facing page. Feed them compost or well-rotted manure twice as often or use twice as much. If you supplement with a standard, balanced fertilizer, spread in spring using only one-fourth the amount recommended for vegetables. If growth isn't lush, give plants another boost in midsummer using an equally low dose.

Herbs that you harvest more than once will benefit from a boost, too. Instead of waiting until midsummer, feed them each time you cut them back to encourage regrowth. Basil, cilantro, arugula, and mints appreciate such treatment. Half-strength fish emulsion (or a fish emulsion/liquid seaweed blend) is a great way to give harvested annual herbs a boost.

Because mints spread, they are easiest to control if given a spot of their own. To grow them among other herbs, provide physical restraint. Cut the bottom off a 5-gallon tub, leaving sides at least 18 inches high. Sink this almost completely into the soil and plant mints inside.

Providing Good Drainage

Good drainage is the most important requirement for herbs. In all but the driest climates and the sandiest soils, they thrive in raised beds. Improve the drainage of heavy soils and the moisture retention of sandy soils with — you guessed it — organic matter. Lean doesn't mean low in organic matter; herbs grow best in humusy loam. Give them an inch of compost or leaf mold every spring after the soil warms to maintain adequate levels of humus.

If your soil is very well drained or tends to be dry, herbs will appreciate organic mulch to keep the soil somewhat moist. In humid climates mulch can encourage rot, so spread it thinly and keep it an inch or two away from plant stems. Where winters are wet, remove organic mulch in fall or replace it with gravel mulch. Cold, wet winter soil kills perennial herbs more often than extreme temperatures. Gravel mulch helps water drain quickly away from the base of plants, reducing humidity and increasing chances of winter survival. Where extremely cold temperatures are also a problem, add a loose winter cover mulch such as evergreen branches or straw.

MASTER GARDENING TIPS

Herbs That Are Moderately Heavy Feeders

These plants like a richer soil than other herbs. Give them more organic matter. Be prepared to supplement this with compost tea or dilute liquid fertilizer if growth isn't as lush as you'd like. Some of these herbs really need just a bit more moisture, which is guaranteed when soils are rich in humus. Even herbs that prefer partial shade can be grown in full sun if given rich, evenly moist soil. (Mints and lemon balm are good examples.)

- Angelicas *(Angelica archangelica, A. atropurpurea)*
- Anise hyssop *(Agastache foeniculum)*
- Arugula *(Eruca vesicaria* subsp. *sativa)*
- Basil *(Ocimum basilicum)*
- Bee balm *(Monarda didyma)*
- Chervil *(Anthriscus cerefolium)*
- Cilantro/coriander *(Coriandrum sativum)*
- Fennel *(Foeniculum vulgare)*
- Fenugreek *(Trigonella foenum-graecum)*
- French sorrel *(Rumex scutatus)*
- Garlic *(Allium sativum)*
- Horseradish *(Armoracia rusticana)*
- Lady's-mantle *(Alchemilla mollis)*
- Lemon balm *(Melissa officinalis)*
- Lemon verbena *(Aloysia triphylla)*
- Lovage *(Levisticum officinale)*
- Orrisroot *(Iris germanica* var. *florentina)*
- Parsley *(Petroselinum crispum)*
- Pennyroyal *(Mentha pulegium)*
- Perilla ("beefsteak plant") *(Perilla frutescens)*
- Saffron *(Crocus sativus)*
- Sweet cicely *(Myrrhis odorata)*
- Sweet woodruff *(Galium odoratum)*
- Valerian *(Valeriana officinalis)*

Growing Perennial Vegetables and Fruits

Crops that last more than one year need a different soil strategy from that of most other vegetables. These include asparagus, rhubarb, and horseradish as well as bush fruits. Some people grow any or all of these at one edge of the vegetable bed. It may be better to grow them in their own spot; then you won't bump into thorns or damage roots while working in your vegetable beds.

Use organic mulch and topdressing or sidedressing to provide a maintenance diet of nutrients and organic matter. If you use long-lasting mulches such as wood chips, you won't have to replenish often. For strawberries, use pine needles or straw so you can eventually dig them into the soil.

Preparing the Soil

Prepare the soil well before planting. Ideally, start a year ahead. Test the soil, amend if needed, and grow a season's worth of green manures to build good tilth and fertility. The buckwheat-rye combination will really reduce your weeding chores the following season. If you can't work that far ahead, try to test soil and adjust pH at least six months before planting. Your plants will grow better and faster.

If you have any reason to suspect heavy soil or poor drainage, double-dig the soil and plant in raised beds. Before planting, dig in lots of organic matter and some rock phosphate to stimulate good root growth and increase frost resistance. Other nutrients can be scratched into the surface now or later.

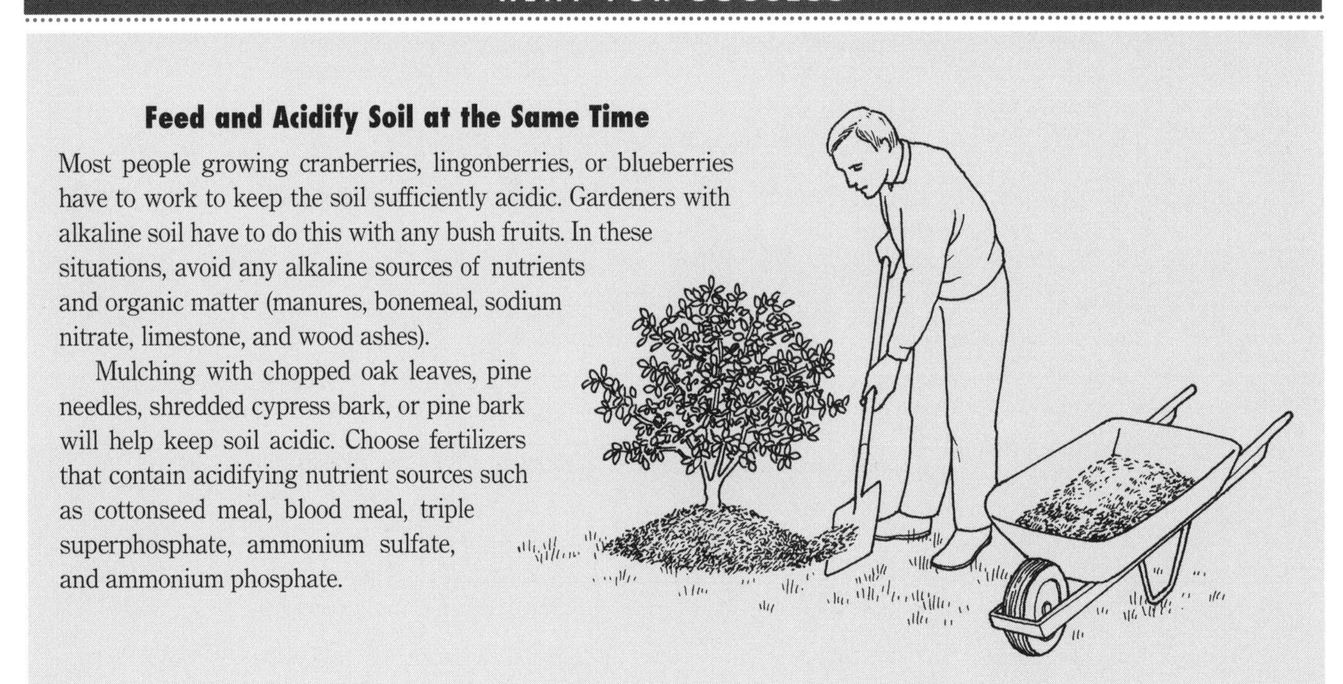

HINT FOR SUCCESS

Feed and Acidify Soil at the Same Time

Most people growing cranberries, lingonberries, or blueberries have to work to keep the soil sufficiently acidic. Gardeners with alkaline soil have to do this with any bush fruits. In these situations, avoid any alkaline sources of nutrients and organic matter (manures, bonemeal, sodium nitrate, limestone, and wood ashes).

Mulching with chopped oak leaves, pine needles, shredded cypress bark, or pine bark will help keep soil acidic. Choose fertilizers that contain acidifying nutrient sources such as cottonseed meal, blood meal, triple superphosphate, ammonium sulfate, and ammonium phosphate.

Fertilizing

All of these plants appreciate a nutrient boost in spring when buds start to swell. (Strawberries are an exception; don't feed them in spring.) Organic fertilizers and slow-release formulations will last all season. Concentrated synthetics may require a second application later in the season, especially in sandy soils that are low in organic matter. Don't use more than the application rates recommended on the label; if in doubt, always err on the low side. If you're spreading compost or rotted manure at the same time, use less fertilizer. Pull mulch aside and spread compost, leaf mold, and/or other nutrient sources directly on the soil surface. Keep fertilizers 6 inches (15 cm) from plant crowns, and broadcast evenly out to the drip line of the outermost branches.

Let the Plants Be Your Guide

It's difficult to give any fertilizer recommendations that make sense for the wide variety of soils and available fertilizers. The only advice that's accurate for all situations is to let your plants tell you what they need. If they're not growing well, slowly increase the amount or frequency of feeding; if growth is too lush, cut back. Feeding with compost is always a good idea, because there's no danger of overfertilizing.

DID YOU KNOW?

Is This Plant Legal?

While the federal ban on growing currants and gooseberries was removed in the 1960s, some states and regions still enforce restrictions. All species of *Ribes* can host a disease that affects white pines (white pine blister rust). You may not want to plant currants or gooseberries if you have white pines growing nearby.

Soil Needs for Bush Fruits and Strawberries

Fruit	Ideal pH	Preferred soil	Tips
Blackberry	5.0–6.0	Well-drained, rich in humus	Mulch well — they're no fun to weed!
Black raspberry	5.0–6.5	Well-drained, rich in humus	Can get incurable diseases from healthy-looking red raspberries, so keep 500 feet (152 m) apart.
Blueberry	4.0–5.5	Moist but well-drained, loamy to sandy	Mulch with pine needles or chopped oak leaves; use a bit more nitrogen than for other bush fruits; shallow-rooted.
Cranberry and lingonberry	4.2–5.0	Very moist, rich in humus	Dig in lots of acidic ground bark or peat moss before planting.
Currant, red	5.5–7.0	Most soils, except poorly drained ones	Relatively heavy feeders; add some nitrogen; shallow-rooted.
Gooseberry	5.0–6.5	Most soils, except poorly drained ones	Add potassium and magnesium but not much nitrogen.
Grape	5.5–7.0	Any soil that's well drained and not compacted	Deep roots need loose, deep soil and subsoil; too much fertilizer or very rich soil causes sour, late grapes.
Red raspberry	5.5–7.0	Well-drained, rich in humus	Mulch well to keep weeds down; surround bed with lawn so you can easily mow off spreading suckers.
Rhubarb	5.5–7.0	Most soils, especially if rich in humus	Heavy feeder; sidedress with compost or aged manure in midsummer and fall.
Strawberry	5.0–6.5	Well-drained	Grow in raised beds; don't feed in spring or you'll get soft berries; fertilize in late summer.

Managing Soils for Vegetables

Vegetables must grow quickly and steadily to be of good quality. That requires evenly moist, well-drained soil with well-balanced nutrients. Only Jerusalem artichoke qualifies as a light feeder; it needs no fertilizer other than compost. The heavy feeders like rich, fertile soil, so they use a bit more fertilizer. The vegetables considered average feeders need less nitrogen but still require lots of organic matter, plus adequate levels of phosphorus and potassium to thrive. Root crops also require loose, deep soil; compacted or rocky soils will make them stunted or deformed. Root crops benefit from double-dug or raised beds.

Preparing the Bed

Fortunately, vegetable gardens are the easiest to supply with nutrients and organic matter. At some point in the season (winter, or between successive crops in mild climates), vegetable beds can be emptied. Then just spread compost or other organic matter and nutrient sources. Empty beds also allow you to grow green manures, so you can raise all the organic matter you need on the spot. With legume green manures, you can even fulfill much of your nitrogen requirements for an entire season. Mulching is another way to maintain organic matter levels; it offers the added benefits of keeping soil moist and reducing weeds.

- **In fall:** Ideally, dig in raw forms of organic matter (chopped leaves and rough, unfinished compost) and rock powders (limestone, rock phosphate) in the fall so that their nutrients will be available by spring.

- **In spring:** Other nutrients, especially nitrogen and fast-acting types, should be added in spring or shortly before planting. Scratch these into the top few inches (cm) of soil; you don't need to dig them in deeply. Unless soil tests show ample to high levels of nitrogen, phosphorus, and potassium, plan on adding each (or a balanced fertilizer) in spring before planting. For many vegetables, especially fast-growing types such as radishes and lettuce, this is enough for the entire season.

- **Anytime:** Compost and well-aged manure can be added anytime and left on the surface as a topdressing or mulch.

MASTER GARDENING TIPS

Choosing a Site for Vegetables

Here's a checklist of characteristics for an ideal site. Few vegetables tolerate shade, so adequate sunlight outweighs other factors.

- At least six hours a day of direct sunlight (more is better).
- Loamy, deep, fertile soil.
- Easy access to water.
- A very slight slope, toward the south if possible, to ensure good drainage and adequate air circulation.
- No low spots that collect standing water.
- In windy areas: shelter from the prevailing wind (with a living windbreak or a fence).

Soil Preferences of Vegetables

The following chart gives ideal soil pH and fertility conditions for various vegetables. Remember that the pH ranges are approximate; many plants will tolerate more extreme conditions (especially if organic matter levels are good). The "Nutrient Requirement" column shows whether plants prefer high fertility (heavy feeders, which may need sidedressing) or average fertility; it also shows which crops shouldn't get extra nitrogen. The family is given to help you group plants for crop rotations.

Vegetable	Preferred pH	Nutrient Requirement	Plant Family
Asparagus	6.0–8.0	Heavy feeder	Lily
Bean (green)	6.0–7.5	Average (low N)	Legume
Beet	6.0–7.5	Heavy feeder	Beet (goosefoot)
Broccoli	6.0–7.0	Heavy feeder	Cabbage (mustard)
Brussels sprout	6.0–7.5	Heavy feeder	Cabbage (mustard)
Cabbage	6.0–7.5	Heavy feeder	Cabbage (mustard)
Cantaloupe/muskmelon	6.0–7.5	Heavy feeder	Cucumber
Carrot	5.5–7.0	Average (low N)	Carrot
Cauliflower	5.5–7.5	Heavy feeder	Cabbage (mustard)
Celery	5.8–7.0	Heavy feeder	Carrot
Chicory	5.0–6.5	Heavy feeder	Daisy (composite)
Corn	5.5–7.5	Heavy feeder	Grass
Cowpea/southern pea	5.0–6.5	Average (low N)	Legume
Cucumber	5.5–7.0	Heavy feeder	Cucumber
Eggplant	5.5–6.5	Heavy feeder	Tomato (nightshade)
Endive/escarole	5.8–7.0	Heavy feeder	Daisy (composite)
Garlic	5.5–8.0	Heavy feeder	Lily
Horseradish	6.0–7.0	Average	Cabbage (mustard)
Jerusalem artichoke	6.5–7.5	Average	Daisy (composite)
Kale/kohlrabi	6.0–7.5	Heavy feeder	Cabbage (mustard)
Leek	6.0–8.0	Heavy feeder	Lily
Lettuce	6.0–7.0	Heavy feeder	Daisy (composite)
Lima bean	6.0–7.0	Heavy feeder (low N)	Legume
Mustard/Oriental greens	6.0–7.5	Heavy feeder	Cabbage (mustard)
Okra	6.0–7.5	Heavy feeder (low N)	Mallow
Onion	6.0–7.0	Average to heavy	Lily
Parsley	5.0–7.0	Heavy feeder	Carrot
Parsnip	5.5–7.0	Average	Carrot
Pea	6.0–7.5	Average	Legume
Pepper	5.5–7.0	Average (low N)	Tomato
Potato	4.8–6.5	Heavy feeder	Tomato (nightshade)
Pumpkin	5.5–7.5	Heavy feeder	Cucumber
Radish	5.5–7.0	Average (low N)	Cabbage (mustard)
Rutabaga	5.5–7.0	Average (low N)	Cabbage (mustard)
Soybean	5.0–7.0	Average (low N)	Legume
Spinach	6.0–7.5	Heavy feeder	Beet (goosefoot)
Squash, summer	6.0–7.5	Heavy feeder	Cucumber
Squash, winter	5.5–7.0	Heavy feeder	Cucumber
Sweet potato	5.5–6.8	Average (low N)	Morning-glory
Swiss chard	6.0–7.5	Heavy feeder	Beet (goosefoot)
Tomato	5.5–7.5	Heavy feeder	Tomato (nightshade)
Turnip	5.5–6.8	Average (low N)	Cabbage (mustard)
Watermelon	5.5–6.5	Heavy feeder	Cucumber

Rotating Vegetable Crops

Rotating crops is easy with advance planning, but it requires record keeping. The only challenging part is figuring out which vegetables to group together. It's really simple if you can devote one bed to each family in a garden of same-size beds. Then you simply move everything over by one bed each year. But few people grow equal amounts from the different plant families, so combining and shuffling are required.

It's easier to manage the garden if late crops are grown together in one spot. That way you can till the rest of the garden and plant winter cover crops without having to dodge kale plants. For early crops such as peas, plan ahead so you won't have to till in spring before planting. Add your annual dose of organic matter in fall and choose oats or annual rye for a cover crop. Either can be raked or pushed aside for early spring planting; they're winter-killed, so you don't have to turn them under to stop spring growth.

HINTS FOR SUCCESS

A Sample Five-Year Plan for Crop Rotation

A: Root crops of any family
B: Nitrogen-building legumes
C: Nitrogen-lovers (corn, cabbage family)
D: Green manures (one slow grower such as alfalfa or a sequence of faster buckwheat and rye)
E: Large heavy feeders (tomatoes, squash)

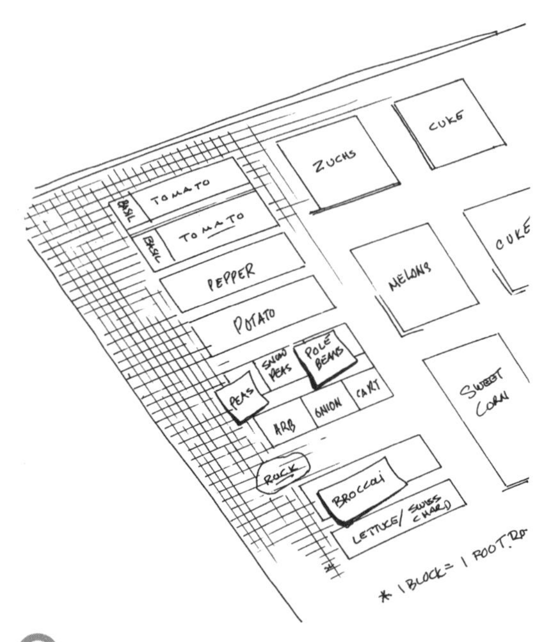

1 Decide which vegetables you want to grow. Make a list of those that fall in the same family. Figure out how much space you want to devote to each vegetable.

2 Arrange these vegetable groups into equal-size blocks or entire beds. If you can't plant a whole block to one family, group heavy feeders together. Root crops work well together, even though they're not in the same family. If a block is devoted mostly to one plant or plant family, ignore the few other plants to simplify things. To save a lot of erasing, try using stick-on tags for different crops.

Getting the Most from Crop Rotations	
Family	**Tips**
Cabbage	Precede with legumes to build soil; use only well-decomposed compost and manure.
Carrot and beet	Best followed by legumes or corn, but do well following any group; don't apply manure the same season or you'll get hairy roots.
Cucumber	Add lots of manure and/or fresh organic matter; plant after winter rye to reduce insects and weeds and to build organic matter; follow with legumes or light feeders (or enrich soil before planting more heavy feeders).
Grass (corn, rye, wheat, etc.)	Alternate with any other group (assuming all residues are returned to the soil); use to soak up nitrogen if too much fertilizer was applied; like lots of manure or fresh compost.
Legumes	Alternate with any other group; can be grown in same spot after 2 years if necessary.
Lily	Good before or after legumes; don't plant after a winter cover crop.
Tomato	Try for at least 4 years before planting bed to another member of this family; follow with legumes or light feeders; precede with rye or rye/vetch combination (see tip on page 108).

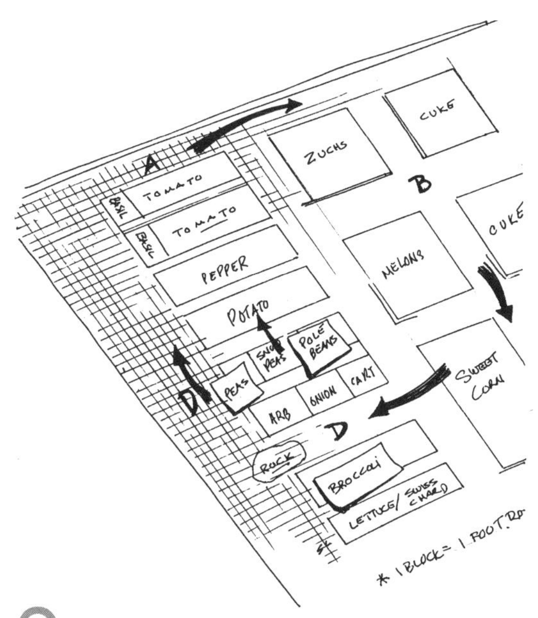

3 Plan your complete cycle of rotation. Ideally, crops in the same family (especially the cabbage family) shouldn't be grown in the same spot more than once every three years. Tomatoes and potatoes benefit from even longer rotations with other members of their family; it's essential where diseases are a problem. Where possible, alternate heavy feeders with light feeders.

DID YOU KNOW?

Benefits of Crop Rotation

- Improves yields (crops consistently yield 10 to 20 percent more in rotations than when grown repeatedly on same soil).
- Improves soil structure; deep-rooted plants help open up channels to improve air and water movement for later crops.
- Maintains fertility by reducing nutrient depletion.
- Uses fertilizers more efficiently.
- Increases diversity of soil organisms.
- Reduces disease problems by robbing pathogens of their hosts.
- Controls insect pests.
- Including annual or winter rye or buckwheat in rotation greatly reduces weeds and builds up organic matter.
- Including legumes in rotation builds up nitrogen levels in soil.

Sidedressing

Heavy feeders appreciate sidedressing partway through the season to supply more nutrients. The best time for their snack varies (see box below). One of the best ways to sidedress is with a couple of handfuls per plant of good compost or aged/dehydrated manure. Alfalfa meal is another organic option; use 1 to 2 tablespoons around each plant and scratch lightly into soil. Foliar feeding with liquid seaweed, a combination seaweed/fish emulsion product, or compost tea supplies micronutrients as well as low levels of major ones. If you're sidedressing with a commercial fertilizer, choose a formula with less nitrogen (6-8-8, 5-10-5, 5-10-10), and scratch it lightly into the soil. If directions for sidedressing say 3 cups per 100 feet of row, that translates to about 1 tablespoon for each large plant (slightly less for medium-size plants such as broccoli).

In sandy or very well-drained soils, sidedressing becomes important even for plants that like average soil. The soil may not be able to hold enough nutrients to carry average feeders through the entire season. Divide the fertilizer recommendation into three or four parts and apply periodically throughout the season.

MASTER GARDENING TIPS

When to Sidedress Heavy Feeders

Asparagus	Twice: before shoots appear in spring and after last harvest (also benefits from manure in fall).
Cabbage family	3 to 4 weeks after planting, or when heads begin to form (don't sidedress turnips and other root types).
Corn	Twice: when about knee-high and when it starts to tassel; likes more nitrogen than other vegetables.
Cucumber family (melons, squash)	When first blossoms appear and/or before vine types start to spread.
Greens (lettuce, endive, Swiss chard, spinach, mesclun mixes)	Feed once or twice before harvesting; for cut-and-come-again types (harvested repeatedly), feed after harvest to encourage regrowth; compost, manure, and fish emulsion work well.
Leeks	When plants reach 6 to 8 inches (15 to 20 cm) tall.
Okra	When first blossoms appear (where growing season is long, they may need a boost a month later).
Onions	Twice: 4 and 6 weeks after planting; use low-nitrogen fertilizer.
Peppers	When first blossoms appear; feed lightly or use low-nitrogen fertilizer or compost (too little phosphorus or too much nitrogen reduces yields).
Pole and lima beans	Give a boost of compost after the first picking (in long-season areas, feed again after another month); don't give these or any other beans added nitrogen or you'll suppress nitrogen fixation.
Potatoes	Just before final hilling, or when buds or first blossoms appear (roughly 6 weeks after planting); don't use raw manure the same season or the previous fall.
Tomatoes	When blossoms or first tiny fruits appear; feeding earlier can delay harvest; too much nitrogen reduces yields.

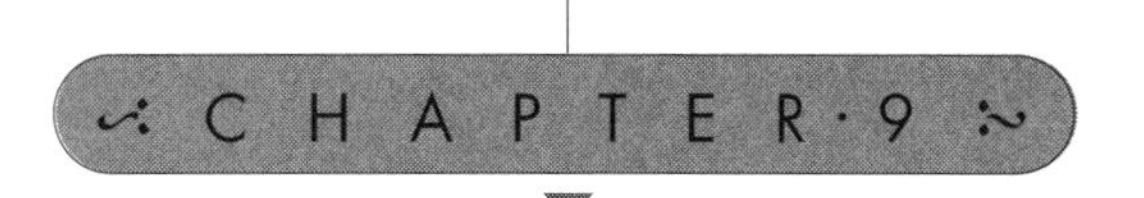

A Soil-Care Calendar

Soil building is an ongoing process. It doesn't stop once you achieve fertile soil — it just gets easier. The difference between creating and maintaining fertile soil is like the difference between losing weight and maintaining your target weight once you've achieved it. Periodic soil tests are like hopping on the scale to see if you're on the right track. If the results don't show what you'd hoped, review the appropriate chapters in this book and modify your approach. Once your soil tests give you the results you want, pat yourself on the back — and keep up the good work!

Keep Records

The easiest way to keep track of improvements is with a garden journal or notebook. You don't need anything fancy; some people find that an old loose-leaf notebook is more flexible than a preprinted volume with lots of pretty pictures. Write down your observations from the tests in chapter 1. A record of what you started with enables you to take credit for any improvements. Eventually, you'll be amazed to look back and see how far your soil has come.

Keep track of soil tests and what amendments or fertilizers you added in response. Make notes about how your garden responded. Yields from the vegetable garden are one of the best ways to measure the effects of soil building. Vegetables respond visibly and rapidly to soil improvements. (Recording yields for different crops also helps you plan how much to grow the following year.)

Make notes about temporary setbacks as well. Pests, diseases, and deficiency symptoms can point to problems missed by soil tests. Record unusual weather; you might find a connection between it and certain pests or diseases. Your observations are perhaps the best single tool for creating fertile soil. Since no two soils are exactly alike, and since you can't test every square foot of a garden, you must look to your plants for information.

The following pages present reminders to help you apply the techniques described throughout this book. They're grouped by season so they make sense for gardeners from Mississippi to Manitoba, from Washington State to Washington, D.C. Modify them as you like to suit your style, your schedule, and your situation.

Fall

• Collect samples for a soil test early enough to have time to add any recommended amendments before the soil freezes. State and private labs aren't as busy in the fall, so you'll get your results back more quickly. Fall is a good time to add calcium (lime) and phosphorus (hard rock phosphate) if your soil test indicates a deficiency of either.

• Fall is the best time to correct acidic soil in lawns and gardens by amending with lime, as indicated by your soil test. (Sulfur works faster, so wait until spring to treat alkaline soils.)

• Every four years, sprinkle granite dust on vegetable beds to supply slow-release micronutrients. Follow application rates on labels.

• As specific crops finish producing, start garden cleanup. Turn under crop residues, chopped leaves, and other sources of organic matter and sow hardy cover crops as you clean out each vegetable bed.

• Rake up the last leaves (or mow with mulching lawn mower) before winter snows.

MASTER GARDENING TIP

Tip for the Season

If fall is always busy for you, put off some cleanup until spring. Spend your limited time on plants that are prone to disease or insects; clean these up first to minimize problems next year. If spring is always busy, be sure to do as much soil preparation as possible in fall.

Dig new garden beds now to get a head start on improving soil. Remove sod and compost it. Plant a cover crop for turning under in spring. If you're too late to plant a cover crop, till in a 4- to 6-inch layer of chopped leaves or compost instead. Or leave sod in place and try the multilayered method of sheet composting instead of tilling.

Topdress lawns with screened compost to reduce thatch buildup. Now is the best time to fertilize lawns (in warm climates, fertilize only cool-season grasses; wait until spring to fertilize warm-season grasses).

Once the growing season ends, or as specific crops finish producing, finish garden cleanup. Destroy any diseased material as well as material containing seeds of obnoxious weeds, or bury well away from the garden. Compost garden debris along with chopped leaves.

Rake leaves for the compost pile, to till into vegetable and new beds or to use as winter mulch (but don't mulch just yet). If you chop with a mower or in a shredder, leaves break down much more quickly. They also make a better mulch; they're less apt to mat down.

Winter

- After the top inch of soil freezes (or at the end of the season in warm climates), spread a thick layer of mulch over bare soil and around perennial plants. (Compost or chopped leaves work well.) Replenish mulch around trees and shrubs, keeping it 6 inches away from trunks to discourage rodents from gnawing bark.
- Clean and oil hand tools before storing for the winter.
- Take an inventory of your garden tools. Toss out what's beyond salvaging and note what needs replacing. Use your inventory to plan the perfect storage shed, or to make improvements to your current system.
- In areas where it's too cold or inconvenient to make frequent trips to the compost pile in winter, start an indoor worm bin.

MASTER GARDENING TIP

Tip for the Season

On warm days in late winter, or when the snow melts, inventory your garden. Make a list of chores for spring. Where the ground doesn't freeze, warm winter days are great times for getting rid of weeds. Try not to step in beds while the soil's still wet.

Where rodents gnaw bark of trees and shrubs each winter, install a protective collar until spring. Cut a 1-foot-wide piece of ½-inch-mesh hardware cloth long enough to wrap around trunk, overlapping a couple of inches. Bend cut ends to hold cylinder in place around trunk.

After the ground freezes in very cold climates, add an additional thick layer of loose mulch over perennials to prevent frost heaving and protect from extreme temperatures. Branches from a discarded Christmas tree work well.

- Add new bedding to existing indoor worm bins if needed.
- If you don't have a map of your garden, make one now before snow covers the ground.
- Make a wish list for yourself of garden tools or other supplies, or a gift list for your gardening friends.
- Order tools, soil amendments, and special fertilizers from garden catalogs when you order your seeds and plants. Order early to avoid disappointment.
- Buy or make a special garden calendar or notebook for yearly records of soil tests, soil improvements undertaken, crop rotations and yields, and your observations.
- Review any notes from last year's garden. Transfer important reminders (such as which crops showed deficiency symptoms) onto a to-do list or calendar for this year. Make plans for the coming season.

Save wood ashes to spread on the garden. Store in a dry spot or waterproof container. You can spread cooled ashes on bare soil or beneath fruit trees and ornamental shrubs even over the snow. Avoid spreading wood ashes around acid-loving plants such as hollies and blueberries.

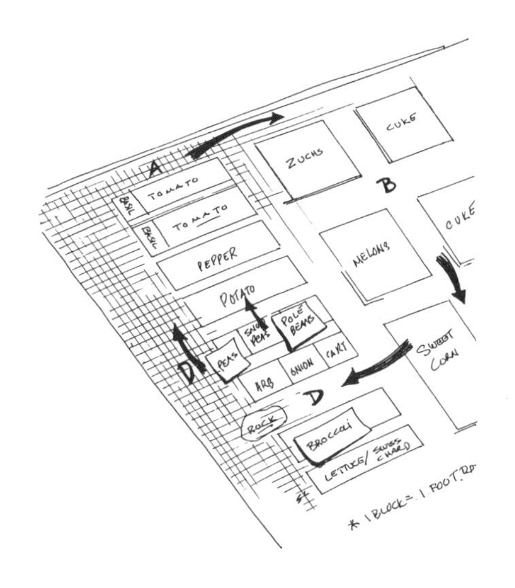

Plan crop rotations to prevent diseases. If you're already rotating your vegetables, review your plan. Note where last year's crops went and decide where this year's should go.

Spring

- If you didn't plant a winter cover crop or dig in chopped leaves last fall, spread compost or well-rotted manure on vegetable beds. Mix into soil as soon as it is dry enough to work.
- If you prepared beds last fall, simply rake the surface smooth before planting.
- Once danger of hard frost has passed (usually when forsythias bloom), pull back mulch from perennials and remove mulch from annual beds to allow soil to warm up.
- Turn leaf piles and compost bins that sat all winter.
- Stop adding kitchen scraps to indoor worm bins a month or so before you want finished compost. This gives worms time to break down the scraps already in the bin. (Add to outdoor bins or piles instead.)

MASTER GARDENING TIP

Tip for the Season

Remember that working wet soil causes compaction and can wreak havoc with soil structure. Don't work your soil too soon in spring. Review "When to Work Your Soil" on page 10 and wait until it's moist, not soggy, to dig in.

As soon as soil dries out enough to work, dig in or till winter cover crops. They need at least a couple of weeks to decompose before you plant; winter rye needs four weeks if you're going to plant seeds.

Start cleaning up the garden whenever weather permits. Remove stray leaves, winter debris, and anything you missed last fall. Start a new compost pile with the debris.

• Test soil if you didn't get around to it last fall.

• Add nutrients recommended by your soil test results. Lime and organic sources of phosphorus are best added in fall; they can be added in spring, but they take a while to become available.

• Spring is the best time to correct alkaline soil by amending with sulfur, as indicated by your soil test.

• After several sets of leaves develop, spread mulch around plants including berry bushes, perennials, and shrubs. If slugs or snails were a problem last year, wait until early summer to mulch — but be prepared to spend more time weeding.

• In warmer climates, fertilize warm-season lawn grasses in mid- to late spring.

• To prevent soil compaction, let soggy, newly thawed lawns dry before driving heavy equipment over them.

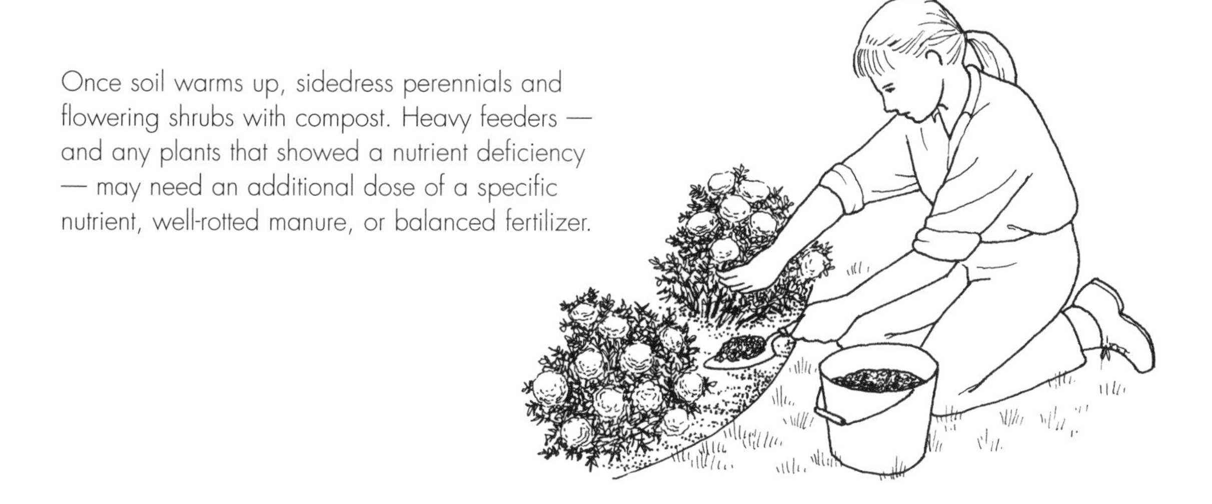

Once soil warms up, sidedress perennials and flowering shrubs with compost. Heavy feeders — and any plants that showed a nutrient deficiency — may need an additional dose of a specific nutrient, well-rotted manure, or balanced fertilizer.

In cooler climates, once lawn grasses start growing rapidly you can apply a spring boost of fertilizer.

Summer

• Inspect plants for any signs of nutrient deficiencies, pests, or diseases. Record any garden problems, and what you did for them.

• Replenish mulch as needed.

• Water gardens and lawns, soaking soil deeply, if rain is scant.

• Plant buckwheat or other green manure crop in unused areas of the garden, or areas that you want to turn into gardens next spring.

• Give heavy-feeder vegetables a midseason nutrient boost. Options include sidedressing with compost or well-rotted manure, and spraying leaves with a liquid seaweed/fish emulsion mix or compost tea.

• In cool climates, stop fertilizing roses and other shrubs or trees in midsummer so plants have time to harden off before frost. This reduces winter damage.

Tip for the Season

Working bone-dry soil can break down structure into dust. Ideally, soil should be slightly moist when you dig. If you can't wait for a summer shower, water your garden well and wait several hours (or overnight) before any major digging project.

• Till under green manures before they form seeds. Sow a second crop of green manure on top. Also sow green manures where early crops were harvested. In areas with short growing seasons, sow winter cover crops in late summer.

• In warm climates, fertilize warm-season grasses again in late summer.

• Pull or hoe weeds before they form seeds. Prompt removal ensures fewer weeds in the future.

• Start new compost piles, or add to existing bins, whenever you have an abundance of garden trimmings or kitchen scraps.

Have leaf-spot diseases been a problem in your vegetable garden? Try spraying plants (especially tomatoes) with compost tea before symptoms appear, or as soon as you see the first signs.

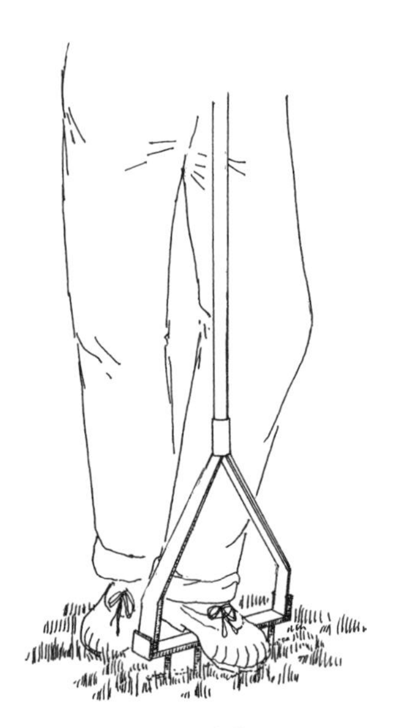

In late summer, prepare soil for new or renovated lawns so that you'll be ready to sow seeds or lay sod in early fall.

Appendixes

Earthworm Suppliers

While you often find earthworms in outdoor compost piles, don't collect these for worm bins. Leave these ordinary earthworms where they are to improve your soil. For worm bins, use only composting worms (red wigglers, *Lumbricus rubellus;* or brandling worms, *Eisenia foetida*). These species eat much more plant material and live happily in the conditions of the bin. In addition to the following national suppliers, you may be able to get composting worms locally; look under "fishing bait" in the Yellow Pages.

Beaver River Associates
P.O. Box 94
West Kingston, RI 02892
(401) 782-8747
Fax: (401) 782-8747
e-mail: riwiggler@aol.com

Gardens Alive!
5100 Schenley Place
Lawrenceburg, IN 47025
(812) 537-8650
Fax: (812) 537-5108

Gardener's Supply Company
128 Intervale Road
Burlington, VT 05401
(800) 863-1700
Fax: (800) 551-6712
Web site: www.gardeners.com

Peaceful Valley Farm Supply
P.O. Box 2209
Grass Valley, CA 95945
(530) 272-4769
Web site: www.groworganic.com

Worm's Way
7850 North Highway 37
Bloomington, IN 47404
(800) 274-9676
Fax: (800) 316-1264
Web site: www.wormway.com

Soil Analysis Laboratories

A & L Agricultural Labs
7621 White Pine Road
Richmond, VA 23237
(804) 743-9401
Fax: (804) 271-6446
Also has laboratories in TN, IN, TX, FL, and Canada. Does not give organic recommendations.

Cook's Consulting
R.D. 2, Box 13
Lowville, NY 13367
(315) 376-3002
Free soil sample kit.

Peaceful Valley Farm Supply
P.O. Box 2209
Grass Valley, CA 95945
(530) 272-4769
Web site: www.groworganic.com

Timberleaf Soil Testing Services
39648 Old Spring Road
Murrieta, CA 92563
(909) 677-7510

Woods End Research Laboratory
P.O. Box 297
Mount Vernon, ME 04352
(207) 293-2457
Fax: (207) 293-2488
e-mail: info@woodsend.org
Web site: www.maine.com/woodsend/

Recognizing Nutrient Deficiencies

To help determine whether a nutrient deficiency is what's ailing your plants, match plant symptoms to those listed for a particular nutrient. The third column here tells you what types of soils and climates are most likely to suffer from each deficiency, offering further clues to your problem. Once you've identified the deficiency, read about how to correct it in chapters 6 and 7. The fourth column here offers tips on how to ensure that the nutrient will continue to be available once levels are restored.

Nutrient	Symptoms of Deficiency	Soils/ Climates Likely to Be Deficient	How to Ensure Availability
Symptoms appear on older, lower leaves first, working up plant:			
Potassium (K)	• Older leaves develop patchy yellow or dead spots; mottling spreads to younger leaves • Leaves normal size but tips and edges look scorched, turn under; may drop • Stems are slender and may fall over	• Sandy soils low in organic matter • Climates with high rainfall or heavy irrigation • Very acidic soils • Soils with too much magnesium and/or calcium (as from overuse of dolomite limestone)	• Maintain good structure • Add well-decomposed organic matter, especially if soil is low in clay • Encourage good root growth
Magnesium (Mg)	• Older leaves splotched with yellow, red, orange, purple, or dead spots but veins remain green • Leaf edges curl upward • Fruits don't taste sweet	• Acidic soils • Sandy soils • Climates with high rainfall or heavy irrigation • Soils with very high calcium or potassium levels	• Maintain ample organic matter • Keep pH above 6 • Avoid overfertilizing • Don't add more lime than needed on acidic soils
Zinc (Zn)	• Older or lower leaves show distinct yellow spots or stripes between veins; spots eventually turn brown and die • Leaves unusually small or narrow with thick, short stalks that make leaves look bunched • Number of flowers and fruits/pods greatly reduced	• Alkaline or overlimed soils • Cold soils • Sandy soils • Soils low in organic matter • Compacted soils • Soils with too much phosphorus, nitrogen, iron, copper, or aluminum	• Keep pH below 7.0 • Maintain abundant organic matter • Minimize soil compaction • Don't overfertilize • Avoid an excess of iron, copper, magnesium, phosphorus, or nitrogen
Molybdenum (Mo)	• Stems are shorter than normal • Leaves thicken and curl (deficiency is difficult to identify — suspect another deficiency unless soil tests low in Mo)	• Sandy soils • Soils low in organic matter • Very acidic soils • Soils with too much copper or sulfur	• Keep pH at 6.3–6.8 • Maintain abundant organic matter • Avoid an excess of other micronutrients
Calcium (Ca)	• New leaves dark and hooked, twisted, or never unfold • Tips of young leaves look scorched • Growing tips may die back • Blossom-end rot of tomatoes, peppers, and melons; other fruits drop off • General lack of plant vigor	• Sandy soils (especially acidic ones near seacoasts) • Sodic (salty) soils • Climates with high rainfall or heavy irrigation • Soils with high levels of aluminum	• Add well-decomposed organic matter, especially if soil is low in clay • Keep soil slightly moist

Nutrient	Symptoms of Deficiency	Soils/ Climates Likely to Be Deficient	How to Ensure Availability
Symptoms appear on entire plant:			
Nitrogen (N)	• Plants fade to pale green or yellow-green • Lowest leaves turn yellow, then brown; usually don't drop • New leaves and shoots smaller than normal • Plants grow too slowly	• Very acidic soils • Soils low in organic matter • Climates with high rainfall or heavy irrigation	• Provide abundant well-decomposed organic matter • Ensure good aeration • Promote good structure • Keep pH at 6.3–6.8
Phosphorus (P)	• Plants look dark or have bluish green cast • Leaves show reddish purple patches on undersides and tips • Lower leaves may yellow, then dry; may drop • Fewer flowers or fruits than normal • Plants grow slowly	• Acidic soils • Cold soils (symptoms may disappear as soil warms)	• Mix applied phosphorus into soil • Keep pH at 6.5–7 • Keep soil slightly moist • Promote biological activity (by providing organic matter)
Symptoms appear on youngest leaves or buds first:			
Sulfur (S)	• Tips and then entire leaves turn yellow • Plants look small and spindly • Plants grow too slowly	• Soils low in organic matter • Very acidic soils	• Maintain ample organic matter
Iron (Fe)	• Chlorosis — younger leaves and buds turn pale green to bright yellow but veins remain green; no dead spots • Overall growth not stunted	• Alkaline or overlimed soils • Poorly drained soils • Soils with too much phosphorus • Soils with high manganese levels	• Keep pH at 6–6.8 • Maintain abundant organic matter • Ensure good aeration • Avoid an excess of manganese (and other micronutrients)
Manganese (Mn)	• Young leaves pale green or yellow around veins, but green areas along veins are wider than in iron deficiency • Small dead spots on leaves	• Alkaline or overlimed soils • Sandy or gravelly soils • Very poorly drained soils • Soils low in organic matter • Soils with too much iron, zinc, or copper	• Keep pH below 7.0 • Maintain abundant organic matter • Ensure good drainage • Avoid overuse of lime • Avoid an excess of iron, zinc, or copper
Copper (Cu)	• Growing tips wilt, eventually die • Young leaves and buds fade to yellow or pale grayish green	• Alkaline or overlimed soils • Sandy or gravelly soils • Peat and muck soils • Soils low in organic matter • Soils with too much nitrogen or zinc • Soils overfertilized with super-phosphate	• Keep pH at 6–6.8 • Maintain abundant organic matter • Ensure good aeration • Don't overfertilize • Avoid an excess of zinc, nitrogen, or phosphorus
Boron (B)	• Leaves twisted, thicker, and purplish to almost black • Flowers drop off plants • Young shoots curl inward and darken, eventually die • Growth is stunted	• Slightly alkaline or very acidic soils • Soils in arid climates • Soils low in organic matter • Soils with too much calcium or potassium	• Keep pH below 7.0 • Maintain abundant organic matter • Avoid overfertilizing • Avoid overuse of lime

Additional Reading

Appelhoff, Mary. *Worms Eat My Garbage.* Kalamazoo, MI: Flower Press, 1982. (One of the first, and still the best, most detailed guide to composting with earthworms.)

Bonnifield, Paul. *The Dust Bowl: Men, Dirt, and Depression.* Albuquerque, NM: University of New Mexico Press, 1979. (A good way to learn about the devastating effects of soil mismanagement in the semiarid Great Plains region of this country in the 1930s. Wind erosion removed vast quantities of topsoil, creating huge dust clouds that darkened the entire eastern seaboard.)

Brady, Nyle C. *The Nature and Properties of Soils.* 8th ed. New York: Macmillian Publishing Co., 1974. (The standard soil science textbook. I much prefer my old, out-of-print edition to the 11th edition published by Prentice-Hall in 1996. The new edition's text is much less readable and the layout is distractingly cluttered.)

Bromfield, Louis. *Malabar Farm.* New York: Harper & Brothers Publishers, 1947. (Bromfield's opinionated account of how he restored his farm's eroded soils to vibrant health; he was one of the founders of modern sustainable agriculture. Reprint editions are available.)

Campbell, Stu. *Let It Rot! The Gardener's Guide to Composting.* Revised Edition. Pownal, VT: Storey Publishing, 1990. (An easy-to-read but thorough guide to composting.)

Campbell, Stu; revised and updated by Donna Moore. *The Mulch Book: A Complete Guide for Gardeners.* Pownal, VT: Storey Publishing, 1991. (Detailed information on types of mulches and their uses.)

Carter, Vernon Gill and Tom Hale. *Topsoil and Civilization.* Norman, OK: University of Oklahoma Press, 1974. (A good book to get you inspired about the importance of controlling soil erosion on a national level.)

Coleman, Eliot. *The New Organic Grower: A Master's Manual of Tools and Techniques for the Home and Market Gardener.* Chelsea, VT: Chelsea Green Publishing Co., 1995. (While aimed at market gardeners, this book contains valuable information for home gardeners, especially on crop rotations.)

Cullen, Mark and Lorraine Johnson. *The Urban/Suburban Composter: The Complete Guide to Backyard, Balcony, and Apartment Composting.* St. Martin's Press, 1993.

DeVault, George, ed. *Return to Pleasant Valley: Louis Bromfield's Best from Malabar Farm.* (A collection of the best of Pulitzer Prize winner Louis Bromfield's essays, many out of print.)

Geiger, Robert L., Jr. *A Chronological History of the Soil Conservation Service and Related Events.* Washington, DC: USDA Soil Conservation Service, Dec. 1955.

Gershuny, Grace. *Start with the Soil.* Emmaus, PA: Rodale Press, 1993. (Basic introduction to soils and soil improvement, with a completely organic focus.)

Hillel, Daniel. *Out of the Earth: Civilization and the Life of the Soil.* Berkeley and Los Angeles, CA: University of California Press, 1991. (Sophisticated, academic discussion of the importance of soils as well as their good and bad management throughout history.)

Hills, Lawrence. *Fertility without Fertilizers.* New York: Universe Books, 1977.

Howard, Sir Albert. *The Soil and Health.* New York: Schocken Books, 1947. (One of the most important early works about organic gardening, composting, and soil improvement.)

Hunter, Beatrice Trum. *Gardening without Poisons.* 2nd ed. Boston, MA: Houghton Mifflin Co., 1971. (Though newer sources are better for pest control, this is still a classic for theory and inspiration.)

King, F. H. *Farmers of Forty Centuries.* (Emmaus, PA: Rodale Press, 1973. (Reprint of 1927 edition; an account of a turn-of-the-century trip through China, Korea, and Japan describing how those countries maintained soil fertility for generations by recycling all organic wastes into the soil.)

Noggle, G. R. and G. J. Fritz. *Introduction to Plant Physiology.* 2nd ed. Englewood Cliffs, NJ: Prentice-Hall, 1983. (An introduction to the biochemistry of plants for gardeners with an interest in science.)

Parnes, Robert. *Fertile Soil: A Grower's Guide to Organic & Inorganic Fertilizers.* 2nd ed. Davis, CA: agAccess, 1990. (Excellent discussion of the importance of organic matter and the roles of different nutrients and fertilizers in creating fertility. Author operates a well-respected private soil testing lab.)

Russell, E.W. *Soil Conditions and Plant Growth.* 10th ed. New York: Wiley, 1974. (Classic work giving a scientific review of soil chemistry with an emphasis on organic matter.)

Stout, Ruth and Richard Clemence. *The Ruth Stout No-Work Garden Book.* Emmaus, PA: Rodale Press, 1971. (An alternative approach to home gardening that substitutes thick blankets of mulch — simplified sheet composting — for tilling the soil.)

United States Department of Agriculture. *Soils and Men: The 1938 Yearbook of Agriculture.* (Still a superb review of building fertility, written in the days before synthetic fertilizers were common. Lots of good information on soil management.)

Waksman, Selman A. *Soil Microbiology.* New York: John Wiley, 1952. (A detailed account of soil microorganisms and their importance to soil health, written by one of the soil scientists who discovered streptomycin.)

Wolman, M. G. and F. G. A. Fournier, eds. *Land Transformation in Agriculture.* New York: John Wiley & Sons, 1987. (Details the many ways in which agriculture has changed the soil and the land.)

Wright, David. *Fruit Trees and the Soil.* London: Faber and Faber, 1960. (Excellent discussion of soil biochemistry and fertility with a focus on tree crops, for home gardeners.)

Useful Conversions

TO CONVERT THESE UNITS	TO THESE UNITS	MULTIPLY BY
inches (in.)	centimeters (cm)	2.54
centimeters	inches	0.394
inches	millimeters (mm)	25.4
millimeters	inches	.0394
feet (ft.)	meters (m)	0.305
meters	feet	3.281
meters	inches	39.37
square inches (sq. in.)	square centimeters (cm^2)	6.452
square centimeters	square inches	0.155
square feet (ft. sq.)	square meters (m^2)	.0929
square meters	square feet	10.76
square yards (sq. yd.)	square meters	0.836
square meters	square yards	1.196
acres	hectares (ha)	.4047
hectares	acres	2.471
cubic yards (cu. yd.)	cubic meters (cm^3)	0.765
cubic meters	cubic yards	1.308
cubic yards	bushels	21.7
bushels	cubic yards	0.046
cubic inches (cu. in.)	cubic centimeters (cm^3)	16.39
cubic centimeters	cubic inches	0.061
cubic centimeters	fluid ounces	0.034
fluid ounces	cubic centimeters	29.57
quarts (qt.)	liters (l)	0.946
liters	quarts	1.057
pounds (lb.)	kilograms (kg)	0.454
kilograms	pounds	2.205
ounces (oz.)	grams (g)	28.35
grams	ounces	0.353
pounds/acre (lb./acre)	kilogram/hectare (kg/ha)	1.12
kilogram/hectare	pounds/acre	0.891
pounds/acre	pounds/100 sq. ft.	.0023

APPROXIMATE WEIGHT TO VOLUME CONVERSIONS	
weight	**approximate volume**
⅙ oz.	1 level teaspoon
½ oz.	1 level tablespoon
8 oz.	1 level cup
1 lb. (16 oz.)	1 pint
2 lbs. (32 oz.)	1 quart
8 lbs.	1 gallon

Index

Italic page numbers refer to illustrations, **bold** page numbers refer to tables

Further Reading

The Big Book of Gardening Secrets, by Charles W. G. Smith. Filled with professional advice for growing the best vegetables, herbs, fruits, and flowers. Teaches beginning and more experienced gardeners how to extend their growing season, grow and use dozens of herbs, care for indoor and outdoor container gardens, cultivate bountiful berry patches and fruit orchards, and grow healthy annuals, perennials, bulbs, and even roses in any climate. 352 pages. Hardcover; ISBN 1-58017-017-X. Paperback; ISBN 1-58017-000-5.

Garden Way's Joy of Gardening, by Dick Raymond. In this best-selling gardening "bible" Dick Raymond shares his proven methods for raised beds, wide rows, and other techniques for creating a bigger harvest with less work. 384 pages with color photos. Paperback. ISBN 0-88266-319-4.

The Gardener's Bug Book: Earth-Safe Insect Control, by Barbara Pleasant. This garden guide shows how to identify and control more than 70 common garden insects using the best homemade and commercial control strategies. 160 pages. Paperback. ISBN 0-88266-609-6.

Let It Rot! The Gardener's Guide to Composting, 3rd edition, by Stu Campbell. A classic guide to the science and art of composting, now completely updated and revised. With delightful humor and wit, Stu Campbell describes the basic life-forms and processes at work in a compost pile, and how to maintain it with organic materials. Includes step-by-step instructions for building compost bins, barrels, and tumblers, a troubleshooting guide, and ideas for using the finished product. 160 pages. Paperback. ISBN 1-58017-023-4.

Pruning Made Easy, by Lewis Hill. Another member of Storey's Gardening Skills Illustrated series. A thorough revision of Lewis Hill's highly acclaimed *Pruning Simplified*, this comprehensive illustrated guide to pruning outlines the many reasons for pruning and explains the different methods used for each purpose. 224 pages. Hardcover; ISBN 1-58017-007-2. Paperback; ISBN 1-58017-006-4.

Seed Sowing and Saving, by Carole B. Turner. Part of Storey's Gardening Skills Illustrated series. A step-by-step illustrated guide to sowing techniqes, the maintenance of seedlings, and seed collecting and storing. Instruction is given for harvesting and saving seeds from more than 100 common vegetables, annuals, perennials, herbs, and wildflowers. 224 pages. Hardcover; ISBN 1-58017-002-1. Paperback; ISBN 1-58017-001-3.